IMAGING IN MOLECULAR DYNAMICS

Charged particle imaging has revolutionized experimental studies of photodissociation and bimolecular collisions. Written in a tutorial style by some of the key practitioners in the field, this book gives a comprehensive account of the technique and describes many of its recent applications.

The book is split into two parts. Part I is intended as a series of tutorials. It explains the basic principles of the experiment and the numerical methods involved in interpreting experimental data. After an historical introduction, this section contains a discussion of the basic experiment and goes on to review image reconstruction techniques. The final chapters discuss alternative approaches to charged particle imaging and explore the possibility of employing the technique in conjunction with femtosecond pump-probe methods to study the excited state dynamics of molecules.

Part II describes a number of different applications. These chapters are more directly research oriented, the aim being to introduce the reader to the possibilities for future experiments. The first chapter describes the power of coincidence imaging detection. The possibilities for probing bimolecular collisions are explored in the following chapters and the final chapter introduces a new approach to ion imaging and velocity mapping called slice imaging.

This comprehensive book will be of primary interest to researchers and graduate students working in chemical and molecular physics who require an overview of the subject as well as ideas for future experiments.

BEN WHITAKER graduated from Sussex University with first class honours in Chemical Physics in 1978 and a D.Phil. three years later with a thesis on state-to-state rotational energy transfer using high-resolution laser spectroscopy under the supervision of Prof. Tony McCaffery. He subsequently worked briefly at the Université de Provence, Marseilles and then at the Université de Paris-Sud where he developed, together with Prof. Philippe Bréchignac, the AON model for rotational energy transfer between small molecules that emphasized the role of angular momentum constraints. In 1983 he returned to Sussex where he continued to work on energy transfer and laser polarization spectroscopy and, with Prof. Tony Stace, on cluster spectroscopy. In 1988 he left Sussex to work with Prof. Paul Houston at Cornell University where he began work on ion imaging. A year later he took up his present position as a Lecturer, and then Reader at Leeds University. He is the author of some seventy papers mainly in the field of molecular reaction dynamics, but his research interests also span a number of areas from computer to combustion science. In his spare time he enjoys making and flying kites, walking in the Yorkshire Dales and the occasional pint of Taylor's Landlord (or Harvey's Best if he is in Sussex). Most of all he enjoys talking science with friends and colleagues, many of whom have contributed to this volume.

IMAGING IN MOLECULAR DYNAMICS

Technology and Applications
(A User's Guide)

Edited by BENJAMIN J. WHITAKER

Department of Chemistry, University of Leeds, Leeds, UK

CAMBRIDGE
UNIVERSITY PRESS

PUBLISHED BY THE PRESS SYNDICATE OF THE UNIVERSITY OF CAMBRIDGE
The Pitt Building, Trumpington Street, Cambridge, United Kingdom

CAMBRIDGE UNIVERSITY PRESS
The Edinburgh Building, Cambridge CB2 2RU, UK
40 West 20th Street, New York, NY 10011-4211, USA
477 Williamstown Road, Port Melbourne, VIC 3207, Australia
Ruiz de Alarcón 13, 28014 Madrid, Spain
Dock House, The Waterfront, Cape Town 8001, South Africa

http://www.cambridge.org

First published 2003

Printed in the United Kingdom at the University Press, Cambridge

Typeface Times 11/14 pt   *System* LATEX 2_ε [TB]

A catalogue record for this book is available from the British Library

Library of Congress Cataloguing in Publication data

Imaging in molecular dynamics : technology and applications : (a user's guide) / edited by Benjamin Whitaker.
p.   cm.
Includes bibliographical references and index.
ISBN 0 521 81059 0
1. Molecular dynamics.   2. Imaging systems in chemistry.   I. Whitaker, Benjamin, 1956–
QD461 .I54   2003
541.3′94 – dc21   2002031561

ISBN 0 521 81059 0 hardback

Contents

Contributors

David W. Chandler
Combustion Research Facility, Sandia National Laboratories, Livermore, California, USA.

Alexei I. Chichinin
Institute of Chemical Kinetics and Combustion, 630090, Novosibirsk, Russia.

Tina S. Einfeld
Institut für Physikalische und Theoretische Chemie, Technische Universität Braunschweig, Hans-Sommer-Str. 10, D38106 Braunschweig, Germany.

André T. J. B. Eppink
Department of Molecular and Laser Physics, University of Nijmegen, Toerooiveld 11, 6525 ED, Nijmegen, The Netherlands.

Karl-Heinz Gericke
Institut für Physikalische und Theoretische Chemie, Technische Universität Braunschweig, Hans-Sommer-Str. 10, D38106 Braunschweig, Germany.

Oded Heber
Department of Particle Physics, Weizmann Institute of Science, Rehovot, 76100, Israel.

Hanspeter Helm
Universität Freiburg, Fakultät für Physik. Hermann-Herder-Str. 3, D-79104, Freiburg, Germany.

Paul L. Houston
Department of Chemistry and Chemical Biology, Cornell University, Ithaca, New York, USA.

Theofanis N. Kitsopoulos
Department of Chemistry, University of Crete, 71409, Heraklion, and Institute of Electronic Structure and Laser, Foundation for Research and Technology-Hellas (IESL-FORTH), 711 10 Heraklion, Crete, Greece.

K. Thomas Lorenz
Lawrence Livermore National Laboratory, Livermore, California. USA.

Christof Maul
Institut für Physikalische und Theoretische Chemie, Technische Universität Braunschweig, Hans-Sommer-Str. 10, D38106 Braunschweig, Germany.

Ulrich Müller
Universität Freiburg, Fakultät für Physik Hermann-Herder-Str. 3, D-79104, Freiburg, Germany.

David H. Parker
Department of Molecular and Laser Physics, University of Nijmegen, Toerooiveld 11, 6525 ED, Nijmegen, The Netherlands.

T. Peter Rakitzis
Department of Physics, University of Crete, 710 03 Heraklion, and Institute of Electronic Structure and Laser, Foundation for Research and Technology-Hellas (IESL-FORTH), 711 10 Heraklion, Crete, Greece.

James L. Springfield
Griffith University, Brisbane, Australia.

Daniel Strasser
Department of Particle Physics, Weizmann Institute of Science, Rehovot, 76100, Israel.

Toshinori Suzuki
Graduate Chemical Dynamics Laboratory, RIKEN, Wako 351-0198, Japan and PRESTO, Japan Science and Technology Corporation, Japan.

Elisabeth A. Wade
Combustion Research Facility, Sandia National Laboratories, Livermore, California, USA. (Current address: Department of Chemistry and Physics, Mills College, Oakland, California, USA.)

Benjamin J. Whitaker
Department of Chemistry, University of Leeds, Leeds LS2 9JT, UK.

Shio-Min Wu
Department of Molecular and Laser Physics, University of Nijmegen, Toerooiveld 11, 6525 ED, Nijmegen, The Netherlands.

Daniel Zajfman
Department of Particle Physics, Weizmann Institute of Science, Rehovot, 76100, Israel.

Preface and acknowledgements

The field of molecular reaction dynamics has made enormous progress since the pioneering experiments of Yuan Lee, Dudley Herschbach and John Polanyi. The intervening years have seen numerous developments in both experimental techniques and theoretical methods. For the authors of this book one of the most exciting of these advances was the introduction of charged particle imaging by Dave Chandler and Paul Houston described in their 1987 paper 'Two-dimensional imaging of state-selected photodissociation products detected by multiphoton ionization' published in the *Journal of Chemical Physics*.

I was extremely fortunate to be able to join Paul Houston in Ithaca in 1988/89 where we constructed the second imaging machine (Dave Chandler's original machine in Sandia having been temporarily put out of action in an unfortunate accident that Paul describes in the first chapter). It was an extraordinarily exciting experience to be involved in those earlier experiments and I am extremely grateful to Paul for the opportunity. The early data showed the power of the technique to provide graphic insight into chemical mechanism but it was difficult to obtain quantitative information because of instrumental problems to do with the arrangement of the ion optics. These were overcome by André Eppink and Dave Parker working in Nijmegen. Their 1997 paper 'Velocity map imaging of ions and electrons using electrostatic lenses: application in photoelectron and photofragment ion imaging of molecular oxygen', which appeared in the *Review of Scientific Instruments*, revolutionized the field and saw a step change in the both the quality of the data that could be obtained and the dynamical insight the experiment could provide. Following the publication of their paper the community of 'imagers' has burgeoned around the world. The advance, described in detail in Chapter 2, was to remove the thin wire meshes in the ion extraction optics that we had all previously striven so hard to get smooth and flat. The problem with using grids to extract the charged particles, however, was that they distorted the trajectories of particles that passed close to the wires and there was a residual image of the grid in the final image, which limited the resolution. In an attempt to overcome this André and Dave designed a new extraction region with many electrodes but still ending in a grid. There is a story that Dave accidentally put his thumb through the grid and that André carried on taking data while the new mesh was on order and so discovered velocity map imaging. The moral being that sometimes it is OK to let your thesis supervisor fiddle with your experiments! The real story was more prosaic; that while waiting for the workshop to get the job done of producing more electrodes, André started with the two open extraction electrodes ('lens') in combination with the repeller plate, but this only goes to show that it's always a good idea to carry on experimenting even when you think things may not be working perfectly.

And so you see also that this book is aimed at students (of all ages) who want to get involved in imaging experiments and do them for themselves. What we have aimed to do is create a 'users guide' to charged particle imaging that not only describes the kind of information that these experiments are capable of revealing but also gives the 'tricks of the trade'. The first part of the book is intended as a series of tutorials. After an historical introduction, in Chapter 2 we describe the basic experiment and the velocity mapping technique in some detail. In Chapter 3 we review the image reconstruction techniques that are used to extract dynamical information from the recorded images. Chapter 4 extends this discussion to cases where the original data is not cylindrically symmetric. Chapter 5 presents an alternative approach to charged particle imaging in which the ratio of intensities in a pair of 2-D images is used to extract timing information. In the process the chapter also reviews recent developments in camera technology and other charged particle imaging detectors. Chapter 6 introduces the idea of being able to use these developments in detector technology to directly record 3-D velocity distributions without the need for the reconstruction algorithms. In Chapter 7 we explore the possibilities of employing charged particle imaging techniques in conjunction with femtosecond pump-probe techniques to probe in particular the excited state dynamics of molecules.

The second half of the book describes a number of applications. These chapters are more directly research orientated, the aim being to introduce you to the possibilities for future experiments and the understanding of molecular dynamics that they will engender. In Chapter 8 the power of coincidence imaging detection is described. In Chapters 9 and 10 the possibilities for probing bimolecular collisions are explored, while Chapter 11 introduces a new approach to ion imaging and velocity mapping called slice imaging.

The idea for the book grew out of the last meeting of the European Funded IMAGINE TMR (training and mobility of researchers) network (ERB 4061 PL 97–0264) that was held at Fodele on Crete, 21–25 October 2000. Incidentally, Fodele is reputed to have been the birthplace of the sixteenth century painter Domenikos

1. Mike Ashfold; 2. Theofanis Kitsopoulos; 3. Thomas Berg; 4. Ben Whitaker; 5. Wim van der Zande; 6. Dave Parker; 7. Peter Andresen; 8. Dave Chandler; 9. Bob Continetti; 10. Toshi Suzuki; 11. Marc Vrakking; 12. Bernd Witzel; 13. Laura Dinu; 14. Ralph Delmdahl; 15. Celine Nicole; 16. Britta Niederjohann; 17. Hanna Reisler; 18. Daniel Strasser; 19. Albert Stolow; 20. Andrew Orr-Ewing; 21. Maurice Janssen; 22. Bernard Bakker; 23. Wolfgang Roth; 24. Hans-Peter Loock; 25. Javier Aoiz; 26. Yan Picard; 27. Danielle Dowek; 28. Dan Neumark; 29. Marco Beckert; 30. Peter Samartzis; 31. Benoit Soep; 32. Hanspeter Helm; 33. Tim Softley; 34. Eloy Wouters; 35. Eckart Wrede; 36. Rolf Bobbenkamp; 37. Steven Stolte; 38. Karl-Heinz Gericke.

Theotocopoulos, otherwise known as El Greco, and whose curious astigmatic perspective is perhaps a fitting backdrop for a book about imaging. Producing a tutorial guide to imaging seemed a good idea at the time (perhaps influenced by the Blue Lagoon and other cocktails served in the bar in the evenings) but it has taken rather longer than anticipated to bring the project to a conclusion.

We (the authors) are grateful to the European Union for their support, either directly as participants in the IMAGINE programme, or through their support of the international network meeting, which allowed us to share the results of the programme with a wider audience. All of the participants at the meeting (see photograph) have contributed to this book, either as authors or through the many discussions we have all shared. The senior scientists involved in IMAGINE were Peter Andresen (Bielefeld), Mike Ashfold (Bristol), Thomas Berg (La Vision, GmbH), Eran Elizur (El-Mul Technologies, Ltd), Theofanis Kitsopoulos (FORTH, Heraklion), Jon Howorth (Photek plc), David Parker (Nijmegen), Isaac Shariv (El-Mul Technologies, Ltd), Wim van der Zande (AMOLF, Amsterdam), and myself, but the real work was done by the young and very talented postgraduate and postdoctoral scientists employed from across the European Union in the various participating laboratories. These were Richard Thomas, Yan Picard, André Eppink, Eckart Wrede, Eloy Wouters, Bernd Witzel, Derek Smith, Bernard Bakker, Christoph Gebhardt, Pablo Quintata, Wolfgang Roth, Hans-Peter Loock, and Ralph Delmdahl. IMAGINE stood for Imaging Network for the Direct Visualisation of Chemical Dynamics. Its inspiration was Lennonesque, and it was the epitome of the European Union's vision for scientific innovation across the member states. Those of us who were involved with it are extremely grateful for the opportunity it offered us to share our ideas and excitement about the potential of imaging methods in chemical dynamics, and the authors hope that this contribution will sow the same seeds of excitement in a new generation of scientists.

All of us would like to dedicate the volume to the memory of Peter Andresen, who very sadly died suddenly soon after. Peter made enormous contribution to molecular dynamics over the years. His experiments on the photodissociation of water were inspirational, and his enthusiasm for 'imaging', from ions to the insides of motor engines, infectious. He is greatly missed.

Abbreviations

ADC	analogue-to-digital convertor
BBO	beta barium borate
CCD	charge coupled device
CFD	constant fraction discriminator
DCS	differential cross-section
DLD	delay-line detector
DPA	differential pre-amplifier
ESDIAD	electron stimulated desorption ion angular distributions
FFT	fast Fourier transform
FTS	frame threshold suppressor
FWHM	full width half maximum
IC	internal conversion
ISC	intersystem crossing
IVR	intramolecular vibrational energy redistribution
KER	kinetic energy release
LIF	laser-induced fluorescence
MCP	micro-channel plate
MF	molecular frame
MSP	micro-sphere plate
OPA	optical parametric amplifier
OPO	optical parametric oscillator
PAD	photoelectron angular distribution
PES	potential energy surface
PR	pulse router
PSA	position sensitive anode
PSD	position sensitive detector
REMPI	resonance enhanced multiphoton ionization

SE	shaping element
TAC	time-to-analogue converter
TDC	time-to-digital converter
TOF	time-of-flight
TOF-MS	time-of-flight mass spectrometry

Part One

Technology

1

Charged particle imaging in chemical dynamics: an historical perspective

PAUL L. HOUSTON

1.1 Introduction

Many problems in molecular dynamics demand the simultaneous measurement of a particle's speed and angular direction; the most demanding require the measurement of this velocity in coincidence with internal energy. Studies of molecular reactions, energy transfer processes, and photodissociation events can be understood completely only if the internal energies and velocities of all products are specified.

Consider the case of a monochromatic photodissociation that produces two fragments A and B. Even if the internal energy distributions of A and B were each measured separately, it would still be necessary to obtain information on their recoil speed in order to determine the internal *energy* of B given a selected level of A. Measurement of the coincident *level* of B would further require that only one parent molecule be dissociated in any particular experiment – a true coincidence experiment. Angular information is also desirable. In photodissociations, for example, the recoil angle with respect to the polarization vector of the dissociating light provides information about the transition moment in the parent molecule and the time-scale of dissociation. Because reactions in molecular beams have many of these same requirements, new techniques for simultaneous measurement of velocity and internal energy are quite important to molecular dynamics.

Many of the current techniques for making simultaneous velocity and internal energy measurements are based on imaging of product molecules or particles with microchannel plate (MCP) detectors. This introductory chapter traces the development of charged particle imaging techniques in molecular dynamics with emphasis on the early ideas that influenced the field. Any such history is, of course, personal and incomplete. This one is no exception. The attempt made here is to provide the background for the exciting work on photodissociation, photoionization, and bimolecular processes using both noncoincidence and coincidence techniques reported in this monograph.

1.2 The need for angular information: vector correlations

From the very beginning of the field of molecular dynamics, angular information has been recognized as essential to the understanding of crossed molecular beam reactions [1]. However, the angular information in photodissociation and photoionization processes has only more recently become of interest. Herschbach and Zare first pointed out that the vector correlation between the parent's transition dipole moment, μ, and the velocity of the product, v, can lead to an anisotropic distribution of photofragments [2]. Photolysis light with a linear polarization defined in the laboratory frame will preferentially excite those parent molecules whose dipole transition moments are aligned parallel to the electric vector of the dissociating light. In the molecular frame, it is nearly always the case that the direction of fragment departure, which defines the recoil velocity vector v, has a fixed angular relationship to μ; for example, in a diatomic parent molecule, v will be either parallel or perpendicular to μ. Consequently, if μ is aligned in the laboratory frame by the dissociating light, v will also be aligned in the laboratory frame, provided that the dissociation takes place rapidly enough so that the alignment of μ is not lost before the moment of fragmentation.

Zare noted that the normalized angular distribution of photofragments should be given by the equation [3]

$$I(\vartheta) = (4\pi)^{-1}[1 + \beta P_2(\cos\vartheta)], \tag{1.1}$$

where ϑ is the angle between v and the electric vector of the dissociating light, $P_2(\cos\vartheta)$ is the second Legendre polynomial, and β is a parameter that describes the degree of anisotropy $(-1 \leq \beta \leq 2)$. Since the dissociation light aligns the transition dipole moment μ in the laboratory frame, and since for a diatomic molecule μ is either parallel or perpendicular to v, measurement of v in the laboratory frame can be used to determine the molecular-frame alignment of μ. Thus, for rapid dissociations of a diatomic molecule, an anisotropy parameter of $\beta = 2$, giving a distribution $I(\vartheta) \propto \cos^2\vartheta$, would indicate that μ is parallel to the breaking bond, while a parameter of $\beta = -1$, giving a distribution of $I(\vartheta) \propto \sin^2\vartheta$, would indicate that μ is perpendicular to the breaking bond. For polyatomic molecules the situation is somewhat more complicated. If one assumes the rapid dissociation limit then $\beta = 2P_2(\cos\alpha)$, where α is the angle between μ and the breaking bond.

Solomon was the first to observe an anisotropic distribution of photofragment recoil velocities [4]. Linearly polarized visible light was used to photolyze bromine or iodine within a hemispheric bulb whose inside surface had been coated with a thin film of tellurium. The pressure of the parent gas was held low enough so that the mean free path of the atomic fragments was larger than the radius of the bulb. The etching rate of the tellurium film as the atoms recoiled to the surface of the bulb was observed to depend on the angle between the velocity and the electric vector of the dissociating light. In both cases, the flux of dissociated atoms peaked perpendicular to the electric vector of the photolysis light, and the experiment was interpreted as indicative of a 'perpendicular' transition with $\beta = -1$. This 'photolysis mapping' method was later used by Solomon *et al.* to investigate the dissociations of aliphatic carbonyl compounds [5]. It was clear, however, that new methods would be needed.

A more quantitative method for measuring recoil velocity distributions was developed in 1969 almost simultaneously by Busch *et al.* [6] and by Diesen *et al.* [7]. In these experiments, a linearly polarized ruby laser was used to dissociate a halogen or interhalogen molecule, while a quadrupole mass spectrometer was employed to detect the arrival time and angular distributions of atomic fragments. A schematic diagram of the apparatus is shown in Fig. 1.1. The principal advantage of this detection technique was that both the direction and magnitude of the recoil velocity could be determined. In the case of I_2, for example, two peaks in the arrival time were observed, corresponding to the two accessible spin-orbit states of the I atom. Even

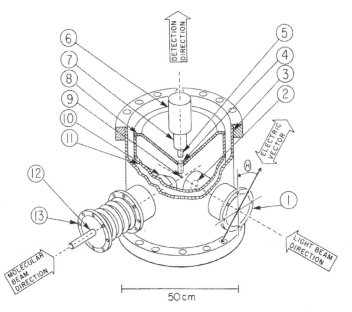

Fig. 1.1. Schematic diagram of photofragment spectrometer. (1) laser beam port, (2) lens, (3) chamber, (4) collimating tube, (5) ionizer, (6) electron multiplier, (7) quadrupole mass spectrometer, (8) partition, (9) interaction volume, (10) collimator, (11) inner wall, (12) molecular beam, (13) molecular beam port. Reprinted with permission from Ref. [8]. Copyright 1972, American Institute of Physics.

though this technique was a great advancement, it did not yet allow determination of the velocity distribution of a *state-selected* product.

1.3 The Doppler effect

Johann Christian Doppler was born in Salzburg in 1803. The effect that bears his name is familiar to all: the frequency of a sound wave heard by an observer depends on the relative speed between the observer and the source of the sound. It was actually Armand Fizeau in 1848 who pointed out that the same effect occurred for light waves. Specifically, the frequency of light absorbed by a moving object is shifted by an amount that depends on the relative velocity between the light source and the object: $\nu_{abs} = \nu_0[1 - (v/c)]$, where ν_0 is the frequency that would be absorbed in the absence of relative motion, v is the relative velocity, and c is the speed of light. Thus, if a molecule to be probed by an absorption-based technique is moving toward the light source, it will absorb at a slightly lower frequency than if it is moving away from the laser. The distribution of absorption frequencies is thus directly related to the distribution of molecular speeds.

The Doppler profile of an absorption can be viewed in a suggestive way: it is simply the projection of the three-dimensional velocity distribution for the probed product onto one dimension, that is, onto the line describing the propagation direction of the probe light. It is well known from spectral broadening studies at low pressure that molecules having a Maxwellian distribution of speeds display a Gaussian spectral line shape whose half-width is proportional to the square root of the temperature. It is less well known that an ensemble of molecules having velocities isotropic in space but characterized by a single speed v exhibits a Doppler profile of equal absorption intensity throughout the line but with sharp edges at frequencies $\nu = \nu_0[1 \pm (v/c)]$. For an arbitrary distribution of speeds, but still for an isotropic distribution in space, the speed distribution $P(v)$ is simply proportional to the derivative of the Doppler profile at the frequency ν corresponding to the velocity v. Although the relationship is less simple if the velocity distribution is not isotropic or if there are angular relationships between the velocity vector and the angular momentum vector [9], even in these cases multiple measurements of the Doppler profile using different detection geometries or different spectral transitions can lead unambiguously to the distribution of speeds. Kinsey demonstrated [10] that the full three-dimensional velocity distribution can be reconstructed by a Fourier transform technique from Doppler profiles measured at several different angles.

From the point of molecular dynamics, use of the Doppler technique had, however, another powerful advantage, state selectivity. Two of the most sensitive laser-based detection techniques are laser-induced fluorescence (LIF) and resonance-enhanced multiphoton ionization (REMPI). Both depend on absorption, and both can be used to detect a product in a single spectroscopic state. By measuring the Doppler profile of products detected by one of these techniques, researchers could finally determine the velocity distribution of a state-selected species. We consider a few important early examples.

Schmiedl *et al.* monitored the Doppler profile of the ground-state H atom produced in the 266-nm dissociation of HI by using vacuum-ultraviolet laser-induced fluorescence [11]. For a fragment whose speed distribution is given measured in the direction **k** making an angle ϑ' with respect to the electric vector of the (linearly polarized) dissociation light, then it can be shown that the Doppler profile for laser-induced fluorescence is given by [11]

$$I(\chi) = (4\pi)^{-1}[1 + \beta P_2(\cos \vartheta')P_2(\cos \chi)], \tag{1.2}$$

where $v_0 \cos \chi$ is the Doppler shift.

Figure 1.2 displays Doppler profiles observed by Schmiedl *et al.* [11] in the 266-nm dissociation of HI for angles $\vartheta' = 0°, 45°$ and $90°$. For photolysis at this wavelength there are two dissociative channels, one producing H + I($^2P_{1/2}$) via a $\Delta\Omega = 0$ (parallel) transition and another producing H + I($^2P_{3/2}$) via a $\Delta\Omega = 1$ (perpendicular) transition. Thus, since the former channel deposits more energy into internal degrees of freedom, we expect to observe a profile composed of slower fragments recoiling with a distribution characteristic of a parallel transition ($\beta = 2$) and faster fragments recoiling with a distribution characteristic of a perpendicular transition ($\beta = -1$). Composite profiles based on (1.2) do indeed fit the data, as shown by the sum of curves a and b in Fig. 1.2.

Another example illustrates the use of the Doppler effect for studies of differential cross-sections in crossed beam collisions. In this experiment, a beam Na($3\,^2P_{1/2}$) was prepared by laser excitation and crossed with a beam of argon atoms [12]. Fine structure changing collisions created Na($3\,^2P_{3/2}$), which was probed by laser excitation with sub-Doppler resolution. Figure 1.3 shows the principle of the technique. The scattered products of the reaction are cylindrically symmetric about the relative velocity vector for the collision and are scattered at an angle ϑ with respect to the original direction of the sodium. By choosing the propagation direction of the probe laser to be parallel to the relative velocity vector for the collision, the authors were able to directly map the differential cross-section as a function of the projection $\cos \vartheta$ on to the Doppler profile, also a function of $\cos \vartheta$. The cross-sections obtained were in quite reasonable agreement with those predicted by theory [13].

A third important early example is the study of the photodissociation of ICN by Nadler *et al.* [14]. These authors used sub-Doppler laser-induced fluorescence to measure the velocity of individual CN internal energy levels. Because more recoil

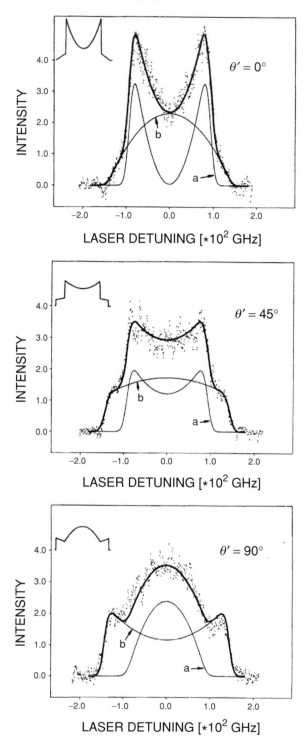

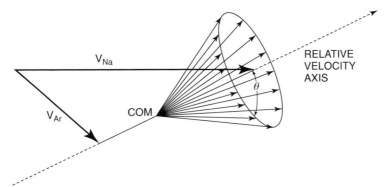

Fig. 1.3. Velocity vector diagram. Bold vectors are initial beam velocities, while light vectors are scattered Na velocities in the center-of-mass frame for scattering through angle ϑ for different values of ϕ. Reprinted with permission from Ref. [12]. Copyright 1978, American Physical Society.

energy is available if the iodine product is produced in the $I(^2P_{3/2})$ ground state than if it is produced in the $I(^2P_{1/2})$ excited state, most Doppler profiles consisted of two components. By resolving these velocity components, the authors were able to determine that high rotational levels of CN were produced in coincidence with ground-state iodine atoms, while low rotational levels are produced in coincidence with excited-state iodine atoms.

The use of a Wiley–McLaren time-of-flight (TOF) spectrometer is an alternative method for making a one-dimensional projection that leads us somewhat closer to charged particle imaging. In an early example, Ogorzalek *et al.* [15, 16] photodissociated CD_3I at 266 nm and ionized the $I(^2P_{1/2})$, $I(^2P_{3/2})$, or $CD_3(v,J)$ products using a second laser. By delaying the pulsed voltage on the repeller and extraction grid, they were able to stretch out the arrival time of the appropriate mass peak. The peaks reveal structure due to the recoil velocity of the I, I*, or CD_3 fragments in just the same way as does a Doppler profile, because products moving originally toward or away from the detector arrive at differing time delays. A variant on the technique sampled a one-dimensional core through the three-dimensional distribution [16]. Ogorzalek *et al.* were able to measure the anisotropic angular distributions of the

Fig. 1.2. (*Left*) Experimental Doppler line profiled (dots) measured at probe angles $\vartheta' = 0°$, 45° and 90°, where ϑ' is the angle between the electric vector of the photolysis light and the propagation direction of the probe light. The theoretical profiles (a) and (b) are for pure parallel ($\beta = 2$) and perpendicular ($\beta = -1$) transitions, weighted such that superposition yields the best fits to the experimental profiles (heavy curves). Inserted in the top left corners are illustrations of the theoretical sum profiles expected under ideal resolution. Reprinted with permission from Ref. [11]. Copyright 1982, Springer-Verlag.

I or CD_3 products and to correlate I/I^* distributions with CD_3 vibrational level. It may have first been a suggestive remark by Ara Apkarian during these experiments that started our own group thinking of how it might be nice to combine the TOF mass spectrometric technique with two-dimensional detection.

In addition to providing correlated information on the products of photodissociation, the one-dimensional velocity projection afforded by the Doppler effect or by TOF mass spectrometry had another important advantage: the measurement was multiplexed in the sense that the velocity of every fragment contributed to some part of the signal. On the other hand, noise on the Doppler profile tended to be amplified in the deconvolution, so that it was sometimes difficult to obtain accurate data. A very reasonable extension of the technique was to make a two-dimensional projection. Fortunately, in the mid-1980s microchannel plate detectors and charge-coupled device (CCD) cameras were just becoming affordable.

1.4 The ESDIAD inspiration

Much of the inspiration for attempting a two-dimensional projection of reaction products came from work in a different area of investigation, surface science. In 1974, Yates and his colleagues first introduced the technique of electron stimulated desorption ion angular distributions (ESDIAD) [17]. These authors discovered that the direction of ions emitted when a surface layer is desorbed by electron bombardment depends on the alignment of the surface layer with respect to the underlying surface. They were able to detect the desorption direction by accelerating the ions with a hemispherical grid into a microchannel plate assembly. Electrons produced in the microchannel plates were further accelerated to a screen and the resulting image was photographed.

A recent example of the sort of results that can be obtained with this technique is shown in Fig. 1.4, which illustrates the results of a study of the rotation of the deuterated methyl group in methyl acetate adsorbed on a Cu(110) surface. At low temperatures, a four-peak pattern is observed that corresponds to a non-rotating methyl group oriented parallel or antiparallel to a particular axis of the copper and having one D atom pointing toward the surface. At higher temperatures, the methyl group rotates freely, and a circular pattern is observed. By fitting patterns at different temperatures to a linear combination of the four-peak and the circular pattern, the authors were able to measure the amount of rotation as a function of temperature, from which they deduced that the barrier height to rotation was 7 ± 4 meV for CD_3 and 12 ± 6 meV for CH_3. The power of images such as these to convey important structural information was an important inspiration leading to the use of similar techniques in product imaging.

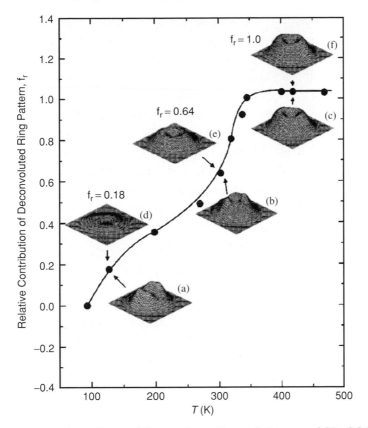

Fig. 1.4. Temperature dependence of the rotation of the methyl group of CD_3COO. Twofold symmetrized D + ESDIAD pattern for acetate/Cu(110) at high coverage. The (a), (b) and (c) patterns are the observed patterns at each temperature. The intensity of the (d), (e) and (f) patterns represent how much the ring pattern contributes in the observed patterns at each temperature. Reprinted from Ref. [18]. Copyright 2000, American Institute of Physics.

1.5 Product imaging

The stage was now set for the development of product imaging. As Dave Chandler remembers it, the idea of combining TOF mass spectrometry with microchannel plate detection of the products came to him while listening to a talk I gave about the experiments of Ogorzalek and Haerri at a November, 1986 Faraday Discussion in Bristol on the dynamics of photofragmentation. Dave actually sketched out an apparatus in his notebook and showed it to me after the talk. I indicated that we had thought somewhat about a two-dimensional projection, but that we certainly didn't have the resources to pursue it. Dave had access to a spare ESDIAD detector, so he proposed that I spend some time at Sandia trying to make a system work. My own recollection is that many of the details of the plan were worked out over beer that

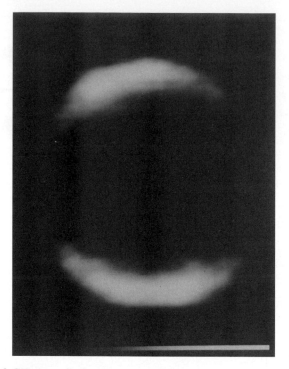

Fig. 1.5. Image of $CH_3(v = 0)$ fragments produced from the photolysis of CH_3I with 266-nm light. The polarization axis of the dissociation light is vertical and in the plane of the figure. Reprinted with permission from Ref. [19]. Copyright 1987, American Institute of Physics.

evening. A Guggenheim Fellowship had provided me the freedom to travel during 1986–87, and the Combustion Research Laboratory at Sandia graciously provided housing for a visit during December.

The first experiment worked very early during the week or so we had dedicated to it. Figure 1.5 shows the first result, an image of the methyl fragment of methyl iodide photodissociation [19]. Although the quality of the image is not outstanding, the fact that we could see the result with our own eyes on the phosphor screen was extremely exciting. These first images were photographed with a normal camera, but we were later to use CCD video cameras to average the results. Several advantages to the technique were immediately obvious. First, the image, though a two-dimensional projection of the full three-dimensional velocity distribution, provided both angular and speed information. Second, because resonance enhanced multiphoton ionization was used to create the ionic products, images could be obtained for state-selected products. For this dissociation wavelength and for the vibrationless level of CH_3, we could see that nearly all of the iodine atoms produced were in their excited spin-orbit state. We were well on our way to studying another molecular dissociation, NO_2, when tragedy struck. The laboratory at Sandia was extremely small and crammed full of equipment. To adjust the laser required (at least for me) standing on a stool and leaning into the center of a large table. Unfortunately, we had located the top of the stool very near to a toggle switch connected to a valve to let in an inert gas in case of vacuum failure. One of us, probably me, must have hit this switch with a toe while adjusting the laser. The next thing we knew, the channelplate had arced and gone dead, bringing us to the end of our experiments that week!

As exciting as the imaging technique was to us, it would never have received wide acceptance had it not been for the efforts of many others. First, the image was a two-dimensional projection of the full three-dimensional velocity distribution. It was possible by using an inverse Abel transform [20–22] to obtain the latter from the former provided the latter had cylindrical symmetry and the former was a projection in a direction perpendicular to the symmetry axis. Several groups helped design stable algorithms for the transform [23–25].

Second, the resolution was still rather limited. Most molecular beams had widths on the order of a few millimetres, and the original imaging apparatus did nothing to prevent the blurring of the image that resulted from dissociations taking place at different locations in the molecular beam. Since the microchannel plates were on the order of 25 mm in radius, the energy and angular resolutions were limited to about 5–10%. A major advance was pointed out by Eppink and Parker [26] who were the first to realize that the incorporation of an Einzel lens in the acceleration region would lead to a great improvement. The lens focuses ions with similar velocities to a common point, regardless of where they originated, and nearly completely eliminates the blurring due to the finite size of the molecular beam/laser overlap.

Another improvement came as the result of work by Chang *et al.* [27], who realized that further increase in resolution could be gained if one carefully analyzed the results of each spot detected by the CCD camera. Under the MCP amplification typical in our own laboratory, each such spot was 5–10 pixels in diameter. By programming a microprocessor to examine each of up to 200 spots per laser shot and to determine the center of the distribution, Chang *et al.* were able to further increase the velocity resolution to the equivalent of one pixel out of the 256-pixel radius of the CCD chip.

Parallel techniques were also being developed at this time. Cold target recoil ion momentum spectroscopy [28] was developed during this period [29–31] to study electron capture and ionization collisions for high-velocity ionization. Electron imaging was also developed, as considered next.

1.6 Electron imaging

Product imaging of positive ions formed by REMPI detection was but one of the areas where charged particle imaging was to become useful. Another area was to the detection of electrons. The first ideas along these lines seem to have an early history [32–34]: Demkov *et al.* [32] were perhaps the first to propose a 'photoionization microscope'. They realized that trajectories of an electron emitted from an atom in different directions may intersect again at a large distance from the atom and create an interference pattern. They proposed building an apparatus to observe the predicted rings. Blondel *et al.* [33,34] eventually realized such a 'microscope' and used it to study the photodetachment of Br^-. It was Helm and co-workers, however, who were the first to create an electron imaging apparatus [35,36]. Figure 1.6 shows a schematic of their apparatus. The instrument is an improvement on previous photoelectron spectrometers in that it provides information on all energies and all angles of the photoelectrons for each shot of the laser.

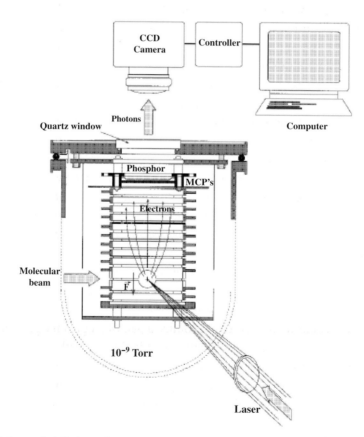

Fig. 1.6. Schematic of photoelectron spectrometer. Reprinted with permission from Ref. [36]. Copyright 1996, American Institute of Physics.

Helm and his coworkers have now used this technique to investigate the ionization of Xe [37–40], Ne [41], H_2 [42], and Ar [39]. In more recent examples, Suzuki [40, 43–45], Haydn [46], and Stolow [46–56] have used femtosecond excitation and ionization to follow excited state dynamics. Many of these applications will be discussed elsewhere in this volume. Two reviews have also appeared recently [57,58].

1.7 Coincidence measurements

Since charged particle imaging has been applied separately to the detection of positive ions and electrons, another reasonable extension is to image the ions and electrons in coincidence. Photoion–photoelectron coincidence methods have been employed fruitfully since the 1970s [51], but a major advance was made by de Bruijn and Los [52], who used a channel plates and a segmented anode to detect coincidence events. The technique was adopted and improved for studies of coincident photofragment detection by Neumark and his group [53]. Somewhat later, Continetti and coworkers pioneered the coincident detection of both the photodetached electrons as well as the subsequent neutral fragments of photodissociation, and several reviews of this work have appeared [59, 60]. A more recent development is the simultaneous imaging using channel plates and CCD cameras of the electrons and photoproducts. For example, Hayden, Continetti, and coworkers have recently investigated the photodissociation of NO_2 by combining photoelectron–photoion coincidence measurements with energy- and angle-resolved imaging [61, 62]. In addition to sorting out the dissociation/ionization mechanism, which turns out to be a three-photon excitation to a repulsive neutral surface followed by a one-photon ionization of the NO, the ability to correlate the direction of the electron with the direction of the ion makes it possible to obtain molecular-frame photoelectron angular distributions [62]. A similar technique has been used by Lafosse *et al.* to study the ionization of NO [63] and O_2 [64].

Eland and his co-workers have developed a different kind of coincidence imaging apparatus [65–67]. One use of the spectrometer is to investigate double ionization. A second is to study the reactions of molecular dications, such as N_2^{2+} and CF_2^{2+}, which react with neutral molecules to produce a pair of singly charged cations. In all cases, the two product cations are imaged in coincidence with a position-sensitive detector that provides the x and y components of their velocities. The z velocity component is derived from the time of flight. For example, in the $CF_2^{2+} + Ar \rightarrow CF_2^+ + Ar^+$ reaction, these authors have correlated the final velocities of the two cations.

A final example includes a triple coincidence experiment performed by Helm and his coworkers on H_3 [68–72]. A single rovibronic state of H_3 was prepared by charge transfer of H_3^+ with cesium and then excited with a laser to effect dissociation into three hydrogen atoms. The hydrogens were monitored by a time- and position-sensitive multihit detector. For each triple hit, the momentum vectors of all three hydrogen fragments were determined. The sixfold differential cross-section of the fragment momentum components in the center-of-mass frame was found to be highly structured and to depend sensitively on the H_3 initial electronic state. Further details can be found elsewhere in this volume as well as in a recent review [73].

1.8 Conclusions

Charged particle imaging has underpinned many new advances in our understanding of atomic and molecular processes. With future development, it is sure to continue to provide new and important data. Coupling of charged particle imaging to synchrotron sources and other femtosecond lasers will also continue to expand its potential. New advances will come on a variety of fronts: increased resolution, as larger MCPs become available; new methods, for example, slicing the three-dimensional distribution rather than projecting it; new applications to other ionization and detachment processes; and deeper understanding, as more complicated processes become uncovered by coincident measurements and other new techniques.

Acknowledgements

The support of the Department of Energy, Office of Basic Energy Sciences under grant DE-FG02-88ER13934 and of the National Science Foundation under grant CHE-9901065 is gratefully acknowledged.

References

1. D. R. Herschbach, *Adv. Chem. Phys.* **10**, 319 (1966); D. R. Herschbach, *Farad. Disc. Chem. Soc.* **55**, 233 (1973); D. A. Case, G. M. McClelland, D. R. Herschbach, *Mol. Phys.* **35**, 541–73 (1978).
2. R. N. Zare, D. R. Herschbach, *Proc. IEEE* **51**, 173 (1963); Zare, R. N., Ph.D. Thesis, Harvard University, Cambridge, MA (1964).
3. R. N. Zare, *Mol. Photochem.* **4**, 1 (1972).
4. J. Solomon, *J. Chem. Phys.* 47, 889 (1967).
5. J. Solomon, C. Jonah, P. Chandra, R. Bersohn, *J. Chem. Phys.* **55**, 1908 (1971).
6. G. E. Busch, R. T. Mahoney, R. I. Morse, K. R. Wilson, *J. Chem. Phys.* **51**, 449, 837 (1969).
7. R. W. Diesen, J. C. Wahr, S. E. Adler, *J. Chem. Phys.* **50**, 3635 (1969).

8. G. E. Busch, K. R. Wilson, *J. Chem. Phys.* **56**, 3626 (1972).
9. G. E. Hall, P. L. Houston, *Ann. Rev. Phys. Chem.* **40**, 375–405 (1989).
10. J. L. Kinsey, *J. Chem. Phys.* **66**, 2560 (1977).
11. R. Schmiedl, H. Dugan, W. Meier, K. H. Welge, *Z. Phys.* **A304**, 137 (1982).
12. W. D. Phillips, J. A. Serri, D. J. Ely, D. E. Pritchard, K. R. Way, J. L. Kinsey, *Phys. Rev. Lett.* **41**, 937–40 (1978).
13. R. P. Saxon, R. E. Olsen, B. Lui, *J. Chem. Phys.* **67**, 2692 (1977).
14. I. Nadler, D. Mahgerefteh, H. Reisler, C. Wittig, *J. Chem. Phys.* **82**, 3885 (1985).
15. G. E. Hall, N. Sivakumar, R. Ogorzalek, G. Chawla, H.-P. Haerri, P. L. Houston, *et al.*, *Disc. Farad. Soc.* **82**, 13 (1986).
16. R. Ogorzalek, G. E. Hall, H.-P. Härri, P. L. Houston, *J. Phys. Chem.* **92**, 5–8 (1988).
17. J. J. Czyzewski, T. E. Madey, J. T. Yates, Jr, *Phys Rev. Lett.* **32**, 777–80 (1974).
18. J.-G. Lee, J. Ahner, D. Mocuta, S. Denev, J. T. Yates, Jr, *J. Chem. Phys.* **112**, 3351 (2000).
19. D. W. Chandler, P. L. Houston, *J. Chem. Phys.* **87**, 1445–7 (1987).
20. D. J. DeRosier, A. Klug, *Nature* **217**, 130 (1968).
21. L. A. Shepp, B. F. Logan, *IEEE Trans. Nucl. Sci.* **NS-21**, 21 (1974).
22. K. R. Castleman, *Digital Image Processing* (Prentice-Hall, Englewood Cliffs, 1979), pp. 184–5.
23. E. W. Hansen, P.-L. Law, *J. Opt. Soc. Am.* **A2**, 510 (1988).
24. L. M. Smith, D. R. Keefer, S. I. Sudharsanan, *J. Quant. Spectrosc. Radiat. Transfer* **39**, 367 (1988).
25. R. N. Strickland, D. W. Chandler, *Appl. Opt.* **30**, 1811 (1991).
26. A. T. J. B. Eppink, D. H. Parker, *Rev. Sci. Instrum.* **68**, 3477–84 (1997).
27. B.-Y. Chang, R. C. Hoetzlein, J. A. Mueller, J. D. Geiser, P. L. Houston, *Rev. Sci. Instrum.* **69**, 1665–70 (1998).
28. M. A. Abdallah, A. Landers, M. Singh, W. Wolff, H. E. Wolf, E. Y. Kamber, *et al.*, *Nucl. Instrum. Meth. Phys. Res.* **B 154**, 73–82 (1999).
29. J. Ullrich, H. Schmidt-Böcking, C. Kelbch, *Nucl. Instrum. Meth.* **A 268**, 216 (1989).
30. J. C. Levin, R. T. Short, C. So, S. B. Elston, J. P. Givvons, I. A. Sellin, H. Schmidt-Böcking, *Phys. Rev.* **A 36**, 1649 (1987).
31. J. P. Grandin, D. Hennecart, X. Husson, D. Lecler, I. Lesteven-Vaisse, D. Lisfi, *Europhys. Lett.* **6**, 683 (1988).
32. Yu. N. Demkov, V. D. Kondratovich, V. N. Ostrovskii, *JETP Lett.* **34**, 403 (1981).
33. C. Blondel. C. Delsart, F. Dulieu, *Phys. Rev. Lett.* **77**, 3755 (1996).
34. C. Blondel, C. Delsart, F. Dulieu, C. Valli, *Eur. Phys. J.* **D5**, 207–16 (1999).
35. H. Helm, N. Bjerre, M. J. Dyer, D. L. Huestis, M. Saeed, *Phys. Rev. Lett.* **70**, 3221 (1993).
36. C. Bordas, F. Paulig, H. Helm, D. L. Heustis, *Rev. Sci. Instrum.* **67**, 2257–68 (1996).
37. C. Bordas, M. J. Dyer, T. A. Fairfield, M. Saeed, H. Helm, *J. Phys. IV* **4**, C4/647–C4/650 (1994).
38. C. Bordas, M. J. Dyer, T. Fairfield, H. Helm, *Phys. Rev. A: At., Mol., Opt. Phys.* **51**, 3726–34 (1995).
39. V. Schyja, T. Lang, H. Helm, *Ultrafast Processes Spectrosc.*, (Proc. 9th Int. Conf., Trieste, 1996), pp. 311–13.
40. V. Schyja, T. Lang, H. Helm, *Phys. Rev. A: At., Mol., Opt. Phys.* **57**, 3692–7 (1998).
41. H. Helm, M. J. Dyer, *Proc. 4th US/Mex. Symp. At. Mol. Phys.*, (World Scientific Publishing, Singapore, 1995), pp. 468–73.
42. C. Bordas, M. J. Dyer, H. Helm, *J. Phys. IV* **4**, C4/691–C4/694 (1994).

43. T. Suzuki, L. Wang, H. Kohguchi, *J. Chem. Phys.* **111**, 4859–61 (1999).
44. L. Wang, H. Kohguchi, T. Suzuki, *Faraday Discuss.* **113**, 37–46 (1999).
45. M. Tsubouchi, B. J. Whitaker, L. Wang, H. Kohguchi, T. Suzuki, *Phys. Rev. Lett.* **86**, 4500–3 (2001).
46. C. C. Hayden, A. Stolow, Non-adiabatic Dynamics Studied by Femtosecond Time-Resolved Photoelectron Spectroscopy. In *Advanced Series in Physical Chemistry Vol. 10: Photoionization and Photodetachment*, ed. C. Y. Ng, (World Scientific, Singapore, 2000).
47. V. Blanchet, A. Stolow, *J. Chem. Phys.* **108**, 4371–4 (1998).
48. S. Lochbrunner, J. J. Larsen, J. P. Shaffer, M. Schmitt, T. Schultz, J. G. Underwood, A. Stolow, *J. Electron Spectrosc. Relat. Phenom.* **112**, 183–198 (2000).
49. M. Schmitt, S. Lochbrunner, J. P. Shaffer, J. J. Larsen, M. Z. Zgierski, A. Stolow, *J. Chem. Phys.* **114**, 1206–13 (2001).
50. V. Blanchet, M. Z. Zgierski, A. Stolow, *J. Chem. Phys.* **114**, 1194–1205 (2001).
51. T. Baer. In *Gas Phase Ion Chemistry*, ed. M. T. Bowers, (Academic Press, New York, 1979); T. Baer, *Adv. Chem. Phys.* **64**, 111–12 (1986).
52. D. P. de Bruijn, J. Los, *Rev. Sci. Instrum.* **53**, 1020–26 (1982).
53. R. E. Continetti, D. R. Cyr, R. B. Metz, D. M. Neumark, *Chem. Phys. Lett.* **182**, 406 (1991).
54. J. P. Shaffer, T. Schultz, M. Schmitt, J. G. Underwood, A. Stolow, *Springer Ser. Chem. Phys.* **66**, 338–40 (2001).
55. S. Lochbrunner, T. Schultz, M. Schmitt, J. P. Shaffer, M. Z. Zgierski, A. Stolow, *J. Chem. Phys.* **114**, 2519–22 (2001).
56. S. Lochbrunner, M. Schmitt, J. P. Shaffer, T. Schultz, A. Stolow, *Springer Ser. Chem. Phys.* **66**, 642–44 (2001).
57. D. M. Neumark, *Ann. Rev. Phys. Chem.* **52**, 255–77 (2001).
58. T. Seideman, Time-Resolved Photoelectron Angular Distributions. Concepts, Applications and Directions, *Ann. Rev. Phys. Chem.* **53**, 41–65 (2002).
59. R. E. Continetti, Dissociative Photodetachment Studies of Transient Molecules by Coincidence Techniques. In *Advanced Series in Physical Chemistry: Photoionization and Photodetachment Vol II*, ed. C. Y. Ng, (World Scientific, Singapore, 2000), pp. 748–808.
60. R. E. Continetti, *Ann. Rev. Phys. Chem.* **52**, 165–192 (2001).
61. J. A. Davies, J. E. LeClaire, R. E. Continetti, C. C. Hayden, *J. Chem. Phys.* **111**, 1 (1999).
62. J. A. Davies, R. E. Continetti, D. W. Chandler, C. C. Hayden, *Phys. Rev. Lett.* **84**, 5983 (2000).
63. A. Lafosse, M. Lebech, J. C. Brenot, P. M. Guyon, O. Jagutzki, L. Spielberger, M. Vervloet, J. C. Houver, D. Dowek, *Phys. Rev. Lett.* **84**, 5987 (2000).
64. A. Lafosse, J. C. Brenot, A. V. Golovin, P. M. Guyon, K. Hoejrup, J. C. Houver, M. Levech, D. Dowek, *J. Chem. Phys.* **114**, 6605 (2001).
65. S. Hsieh, J. H. D. Eland, *J. Phys. B - Atom, Molec. Opt. Phys.* **29**, 5795 (1996).
66. S. Hseih, J. H. D. Eland, *Rapid Commun. Mass. Spectr.* **9**, 1261 (1995).
67. J. H. D. Eland, *Meas. Sci. Technol.* **5**, 1501 (1994).
68. U. Muller, T. Eckert, M. Braun, H. Helm, *Phys. Rev. Lett.* **83**, 2718–21 (1999).
69. R. Reichle, I. Mistrik, U. Muller, H. Helm, *Phys. Rev. A: At., Mol., Opt. Phys.* **60**, 3929–42 (1999).
70. U. Muller, M. Beckert, M. Braun, H. Helm, *Dissociative Recomb.: Theory, Exp. Appl., Proc. IV, Conf.* (World Scientific Publishing, Singapore, 2000), pp. 81–90.

71. U. Muller, R. Reichle, I. Mistrik, H. Helm, J. A. Stephens, M. Jungen, *Dissociative Recomb.: Theory, Exp. Appl., Proc. IV, Conf.*, (World Scientific Publishing, Singapore, 2000), pp. 293–4.
72. I. Mistrik, R. Reichle, H. Helm, U. Muller, *Phys. Rev. A: At., Mol., Opt. Phys.* **63**, 042711/1–042711/10 (2001).
73. C. C. Hayden, *Ann. Rev. Phys. Chem.* **54**, (2003) (to be published).

2

Velocity map imaging: applications in molecular dynamics and experimental aspects

DAVID H. PARKER AND ANDRÉ T. J. B. EPPINK

2.1 Introduction

This chapter aims to introduce you to the practical aspects of molecular dynamics research using imaging methods. Imaging is a rapidly advancing experimental technique full of possibilities. This puts you in position to make unique and important contributions to the field of reaction dynamics, and you will be the first person to *see* the secrets of nature appear on your camera screen. Every scientist lives in part for this exciting result and velocity map imaging more than any other method presents the full picture in living color!

Velocity map imaging is the present day variant of the ion imaging method invented by David Chandler and Paul Houston in 1987 [1]. We discovered the advantages of a simple electrostatic lens for the ion imaging method in 1997 [2]. The improvement was so dramatic that David Chandler convinced us to give it the new name, velocity map imaging. Undoubtedly, you or some other clever scientist will discover a new trick to make imaging work even better in the future. Imaging has much to offer in present-day molecular dynamics research, as illustrated in this chapter. This introduction will lead to the following chapters in this book on experimental aspects, data analysis, angular momentum theory, photoionization, and alternative methods.

2.1.1 Outline of this chapter

The story begins with the all-important Newton spheres and what they tell us about unimolecular and bimolecular processes such as photodissociation and reactive scattering. Although imaging is unique in its direct measurement of the Newton spheres, it is certainly not the only experimental method in this research field – the main competition is with time-of-flight methods. A comparison of the two methods for bimolecular scattering will be made. We will then focus on imaging of photodissociation processes by following the laboratory procedure – first by creating the spheres, then converting them into ions, converting the ions to a two-dimensional image, and converting the image back to the Newton sphere.

2.2 Newton spheres: their concept, creation and analysis

The goal of molecular dynamics research [3] is to develop a complete understanding of a selected physical or chemical event, be it photoionization, photolysis, or the chemical reaction of a given molecule. This requires a detailed knowledge of the potential energy surface for the process, and the dynamics of motion on that surface. By experimentally specifying molecular properties such as laboratory velocity, internal temperature, axis or rotational angular momentum alignment, etc. *before* the event takes place, and fully characterizing the final products *afterwards*, and with the help of a few simplified models, we can usually begin to make sense of the process. Using ultra-short laser pulses, electron imaging, and other approaches, the transformation from reactant to product can sometimes even be probed *during* the process, as discussed by Suzuki and Whitaker in Chapter 7. In this chapter we will focus on the connection between the 'before' and 'after' stage of molecular dynamics processes.

Most processes of interest are simple two-body events that end with the particles departing from each other with a fixed amount of kinetic and internal energy. This includes:

chemical reactions:

$$A + BC \rightarrow ABC^* \rightarrow AB + C,$$

inelastic scattering:

$$AB(v, J) + C \rightarrow ABC^* \rightarrow AB(v', J') + C,$$

photoionization:

$$AB + h\nu \rightarrow AB^* \rightarrow AB^+ + e^-,$$

and photodissociation:

$$AB + h\nu \rightarrow AB^* \rightarrow A + B,$$

where $AB(C)^*$ is a collision or photo-excited complex. Details of the partitioning between kinetic and internal energy, and the type (rotational, vibrational, or electronic) of internal energy can be very informative about how the scattering event takes place.

Let us consider the dissociation process of the excited complex, $AB^* \rightarrow A + B + TKER$, where TKER (Total Kinetic Energy Release) is the total excess energy left over after subtracting the internal energy of the A and B products. Conservation of momentum and energy results in the kinetic energy partitioning

$KER_A = (M_B/M_{AB}) \times TKER$ and $KER_B = (M_A/M_{AB}) \times TKER$. In the following text the factor (M_A/M_{AB}) or (M_B/M_{AB}) is called the mass partitioning factor. For photoionization $M_{e^-} \ll M_{M^+}$, thus the photoelectron receives essentially all TKER, whereas for photodissociation of a homonuclear diatomic such as O_2, TKER is equally shared between the two product O atoms.

Each photodissociation or photoionization event yields two partner fragments flying with equal momentum in opposite directions in the centre-of-mass frame. Repeating the same event for many times, the fragments build up spherical distributions in velocity space. *These are the so-called Newton (velocity) spheres for the process.* The size of the Newton sphere tells us about the balance of internal and translational energy in the reaction. Equally interesting is the surface pattern of the Newton sphere. For the moment we will assume this pattern is anisotropic, i.e., *not* homogeneously distributed across the surface.

2.2.1 Ideal conditions for creating Newton spheres

Laser photodissociation $AB + h\nu \rightarrow A + B$ creates the Newton spheres shown schematically in Fig. 2.1, where mass A < mass B. The first dissociation event (Fig. 2.1a) sends fragment A upward and B downward, while event 2 sends the two again in opposite directions but now almost perpendicular to the vertical axis. If we could somehow arrange that all events 1, 2, 3, etc., start with the parent molecule AB located at the same point in space (origin $x,y,z = 0,0,0$; $\Delta x, \Delta y, \Delta z = 0,0,0$) and with zero initial velocity ($v_x, v_y, v_z = 0,0,0$; $\Delta v_x, \Delta v_y, \Delta v_z = 0,0,0$), then each dissociation event will yield particles with identical speed but varying directions. When we measure the particle's position a fixed time after the event started, then

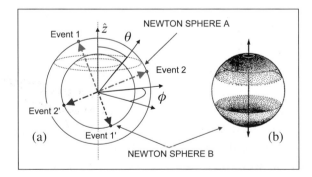

Fig. 2.1. (a) A pair of Newton spheres with spherical coordinates r (not shown), θ, and ϕ, where θ is the polar angle defined with respect to the z axis (vertical axis in the figure), ϕ is the azimuthal angle, and r is the sphere radius. Two events are shown in (a) with equal and opposite momentum. (b) By summing up a large number of events (here for particle B), a surface pattern emerges as shown. Most of the surface intensity of this example is at the poles, representative of a typical $\cos^2\theta$ distribution.

this final (x,y,z) position is directly proportional to the Newton sphere *velocity* (a vector quantity). The *size* of the Newton sphere is directly proportional to the fragment's *speed* (a scalar quantity). Every particle A falls somewhere on the Newton velocity sphere A, and the same applies for B. When we add up the measurements of the A and B velocities for a large number of events (Fig. 2.1b), any pattern that might appear on sphere A is identical to the pattern observed for Newton sphere B.

2.2.2 Consequences of energy and momentum conservation

Conservation of energy and momentum relates the information (radial and angular) on the Newton sphere for one species with that of its partner. Let's consider the photodissociation process $AB + h\nu \rightarrow A + B$ with A having only one possible internal quantum state while B has three internal quantum states i, j, k, equally spaced in increasing energy. Let the internal energy of state i $= 0$ and state k equal to 2/3 of the total energy available, which is just the difference in photon energy $h\nu$ and the bond energy of AB. Suppose that the reaction produces states B in the fractional yield (i : j : k) $= (0.6 : 0.3 : 0.1)$ and, for ease of illustration, that the anisotropy differs greatly for each B quantum state. Reaction producing state i, j, and k have $\cos^2\theta$ (polar), isotropic, and $\sin^2\theta$ (equatorial) angular distributions, respectively.

It is much easier to draw these Newton spheres in two dimensions instead of three as in Fig. 2.1. If we were to 'squash' Fig. 2.1b onto a vertical plane, it would look like a disc with most of the events appearing on the outside edges at the top and bottom of the disc. Velocity map imaging works just in this way – to record the three-dimensional (3-D) Newton sphere we project it onto a two-dimensional (2-D) surface where it appears as a partially filled-in circle. Newton spheres for products A and B(j) are shown in Fig. 2.2. Mass B is set at four times mass A, thus from momentum conservation the radii of corresponding rings appear in image A twice as large as in image B.

Detection of species B in the single quantum state j reveals only one (isotropic) Newton sphere, which corresponds to the middle Newton sphere of A. Detection of species A, however, reveals all three Newton spheres of B corresponding to states i, j, and k, each with their different surface patterns. These spheres are nested. Thus, when crushed, the discs of the smaller sphere fall on the anisotropic inner regions of larger discs. Our analysis program will have to pull these features apart.

2.2.3 Why is the surface pattern anisotropic?

Random gas-phase collisions create Newton spheres with homogeneous (isotropic) surface patterns. An anisotropic surface pattern is the result of a selected

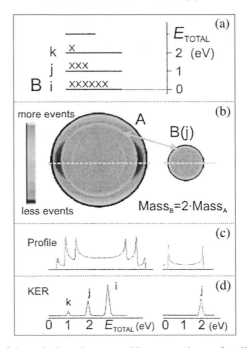

Fig. 2.2. Illustration of the relations between Newton spheres for dissociation of AB with total excess energy E_{TOTAL}. A has only one internal state, B has three states with energy position and populations indicated in panel (a). A 2-D crushed view of Newton spheres for A and state 'j' of B is shown in panel (b). Note the disc size is proportional to speed and not energy. Panels (c) and (d) shows speed profiles for a horizontal line through the 2-D crushed spheres, and the corresponding kinetic energy distributions, respectively.

directionality in the process. This can be introduced experimentally by restricting a degree of freedom of the reactants. The surface pattern of the Newton sphere shown in Fig. 2.1b can arise from *photoselection* due to the polarization vector of the incident photon beam we use for photodissociation. We can induce directionality in several ways, the most common way is to form a molecular beam and cross it at a specified angle with another molecular beam or with a linearly polarized laser beam. In addition, we can also select reactant molecular angular momentum alignment and even molecular axis alignment and/or orientation in more sophisticated experiments. Even fairly unsophisticated beam scattering experiments that merely define a reference axis or plane associated with the collision process can reveal directional properties of the product recoil velocity, angular momentum, molecular axis polarization, etc. These directional or *vector* properties are often interconnected via two-, three-, and possibly higher-order vector correlations. Vector correlations provide very sensitive probes of potential energy surfaces and collision dynamics. The radial *and* angular patterns of the Newton sphere and their correlations with other product properties are thus of great interest.

2.2.4 Experimental requirements

How do we accurately capture a Newton sphere such as the simple spheres shown in Figs. 2.1b and 2.2b? A well-defined initial position *and* velocity of the ABC* complex is necessary in order to obtain the best resolution possible with imaging. While the *(x,y,z)* position of ABC*, i.e., the origin of the Newton spheres, is simply set as the zero point in the experimental (lab) frame, a spread in the origin, $\Delta x, \Delta y, \Delta z$ introduces blurring (non-superposition) of the image. In most reaction experiments an initial AB velocity (v_x, v_y, v_z) can be compensated for, but again more important is the velocity spread $(\Delta v_x, \Delta v_y, \Delta v_z)$, which is determined by the speed ratios $(\Delta v/v)$ of the reactant molecular beam(s). Velocity map imaging, the method of this chapter, uses a wonderful trick with electrostatic lenses to allow capture of the Newton sphere information essentially independent of the origin position spread $\Delta x, \Delta y, \Delta z$. Velocity spreads, however, must be kept as small as possible.

Another requirement in obtaining a meaningful image is that nothing disturbs the fragment velocities during the measurement, meaning that the experiment must be carried out under *single collision conditions*. After the initial collision event the rapidly moving ionic Newton sphere must be left undisturbed by additional collisions for at least 100 μsec ($\sim$10 times longer than the ion flight time). This requires a background pressure of less than 10^{-5} mbar. A high-vacuum molecular beam apparatus is thus necessary.

2.2.5 Bimolecular scattering

It is easy, conceptually and experimentally, to visualize photodissociation as fragments ejected from a parent molecule frozen in space. Photodissociation is a subset of more general two-body scattering process where the momentum of one reactant (the photon) is negligible. In this case the origin of the Newton spheres of the products moves in the centre-of-mass frame of the system in the same direction as the reactant molecule beam. If we look into the molecular beam source, the parent molecule indeed appears to be standing still on the time scale of sphere expansion. This makes a direct mapping of the product centre-of-mass motion in the laboratory frame possible. In addition, the polarization of the photon source is an easily manipulated parameter for inducing directionality in the experiment.

More general scattering processes such as inelastic and reactive scattering can be much more complicated than photodissociation, depending especially on how the products are measured. The complexity will become apparent in Section 2.3 for the standard type of measurement using time-of-flight. In bimolecular scattering the centre-of-mass and relative velocity vectors are constructed with the aid of a so-called 'Newton diagram'. The crossing volume of the two reactant beams defines the origin of the Newton sphere, which moves in the laboratory frame with the centre-of-mass velocity. Using crossed beams, the spatial distribution of angular momentum and axial vectors is not isotropic and is thus said to be polarized. This polarization can result in Newton spheres with anisotropic surface patterns. Just as there is axial symmetry about the polarization vector in photodissociation, there is axial symmetry about the relative velocity vector of two collision partners.

Consider the bimolecular scattering of metastable neon atoms with argon atoms [4]. Associative ionization creates $NeAr^+$ in the process $Ne^* + Ar \rightarrow NeAr^* \rightarrow NeAr^+ + e^-$ and Ar^+ is created by the Penning ionization $Ne^* + Ar \rightarrow NeAr^* \rightarrow Ne + Ar^+ + e^-$. The centre-of-mass velocity vector for this system is shown along with the reactant relative velocity vectors in Fig. 2.3. Because of the mass partitioning factor the electron carries off the majority of TKER. In this geometry, which is typical for crossed-beam scattering, the initial laboratory velocity results in a displacement of the Newton spheres away from the centre of the apparatus. The Newton spheres and their surface patterns for the associative and Penning ionization channels shown in Fig. 2.3 are discussed in detail in Delmdahl *et al.* [4].

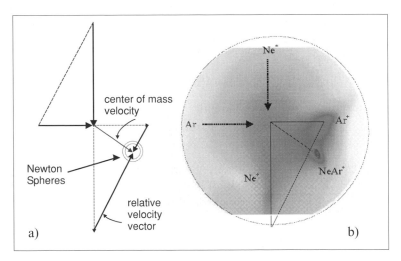

Fig. 2.3. Crossed-beam scattering geometry and a velocity map image of the reaction of metastable Ne* atoms with Ar to form $NeAr^+$ and Ar^+. Velocity vectors are drawn for Ar^+ arrival times. Light from the discharge used to create Ne* also ionizes the Ar and Ne reactant beams, making the two reactant beam velocities visible. Lab frame velocity vectors are shown for Ne* and Ar along with the centre-of-mass and relative velocity vectors. Note that these vectors are determined by momentum balance and are not necessarily perpendicular. Because imaging maps $KE^{1/2}$, the image and Newton (velocity) diagrams are not superimposed.

2.2.6 Characterization of the surface patterns of Newton spheres

2.2.6.1 Photodissociation, photoionization and bimolecular scattering

What can be extracted from the surface pattern of the Newton sphere? For photodissociation, the directionality of the ejected A and B fragments yields information on the direction of the transition dipole with respect to the bond-breaking axis and thus the nature of the excited electronic state. For example, fast photodissociation of a diatomic molecule with a Σ symmetry initial electronic state and a Σ final state results in a parallel $(\cos^2\theta)$ fragment angular distribution, while a Π final state results in a perpendicular $(\sin^2\theta)$ fragment distribution. Mixed transitions and the qualification of 'fast' are discussed later in this chapter.

For photoionization the surface pattern of the Newton sphere reveals the nature of the ejected electron waves [5]. These are coherent sums of the possible s, p, d, f . . . electron wave functions. For small atoms it is possible to predict the photoelectron angular distributions to high accuracy but for molecules this is an active field of research. The kinetic energy information from photoelectron measurements can be very useful in identifying the origin and internal excitation of the partner cations formed in photoionization processes.

The measurement of the surface pattern of the Newton sphere for bimolecular scattering is equivalent to measuring the differential (solid angle) scattering cross-section [3]. Scattering angular distributions are sensitive probes of the intermolecular potential energy surface. For elastic scattering the balance of long-range attractive and short-range repulsive forces as a function of the impact parameter leads to glory and rainbow scattering and quantum oscillations about these angles. Inelastic scattering can introduce more complicated interference effects and rotational rainbows. Angular distributions for reactive scattering also serve as probes of the collision dynamics and reveal information on the intermolecular potential. Here the concepts of forward, backward and sideways scattering for direct reactions and long-lived collision complexes are part of the vocabulary of molecular dynamics. For each type of event, the scattering information is imprinted on the surface pattern of the Newton sphere.

2.2.6.2 Vector notation

During photoexcitation of molecules in a molecular beam the randomly oriented molecules whose transition dipoles happen to lie parallel to the polarization vector ε_p of the light source (z axis in Fig. 2.1) are preferentially excited. Thus, ε_p selects μ, the transition dipole moment, in the laboratory frame. There is an axis of cylindrical symmetry about z – the absorption depends on θ but not on ϕ the azimuthal angle. Bond breaking can occur, and the fragments recoil along the direction of the bond

Table 2.1. *Vector correlations*

Correlation	Name	Symbol	Types
$(\boldsymbol{\mu}, \mathbf{k}')$	Photofragment angular distribution	$\beta\ (-1 \le \beta \le 2)$	perpendicular, parallel, mixed, coherent
$(\boldsymbol{\mu}, \mathbf{j}')$	Photofragment alignment	$A_0^{(2)}$	perpendicular, parallel, mixed ...
$(\boldsymbol{\mu}, \mathbf{j}')$	Product alignment	$A_0^{(2)}$	perpendicular, parallel, mixed ...
$(\mathbf{j}', \mathbf{k}')$	Helicity		clockwise, counter-clockwise perpendicular, parallel ...
$(\mathbf{k}, \mathbf{k}')$	Reaction angular distribution		forward, backward, sideways ...

axis $\mathbf{r}$, which defines the relative recoil vector $\mathbf{k}'$. For a diatomic molecule if $\boldsymbol{\mu}$ is parallel to the bond axis there will be a proclivity for fragments at the poles of the sphere, i.e., along the vertical axis, as depicted in Fig. 2.1. The correlation of $\boldsymbol{\mu}$ and $\mathbf{k}'$ is thus imprinted on the surface of the Newton sphere. If photodissociation is slow compared to molecular rotation the directional correlation of $\boldsymbol{\mu}$ with $\mathbf{k}'$ will be lowered. Even if dissociation is so slow that the dissociation axis is isotropic in the lab frame, the internal angular momentum, $\mathbf{j}'$, in each fragment can still be aligned along the dissociation axis $\boldsymbol{\mu}$ (another two-vector correlation) and thus in the LAB frame along the z axis. Table 2.1 lists these main correlations for photodissociation and bimolecular scattering processes along with the parameters used to characterize them. More details on the β parameter are given in Section 2.4.1

2.2.7 Three-vector correlations

Directionality is a universal property of chemical reactions. A familiar directional property is the steric effect, which for the prototype reaction $AB + C \rightarrow A + BC$ implies that collisions of C with the B end of AB are more reactive than collisions with the A end. We can quantify this by, for example, the total reactivity as a function of the angle of attack on AB by C, i.e., the reactivity as a function of the orientation of the $\mathbf{r}$ vector lying along the A–B axis. It is actually possible to experimentally control the orientation of $\mathbf{r}$ of a polar symmetric molecule in the lab frame using electrostatic fields [6]. We can then ask how $\mathbf{r}'$, the axis of BC, points as a function of $\mathbf{r}$, and even how the whole picture varies as a function of the product scattering angle $\mathbf{k}'$. Such a detailed and revealing question is an example of a three-vector correlation $(\mathbf{r}, \mathbf{r}', \mathbf{k}')$. The most powerful characteristic of imaging is its ability to reveal directly these high-order vector correlations for molecular dynamics processes. We have already introduced two-vector correlations in Table 2.1, especially

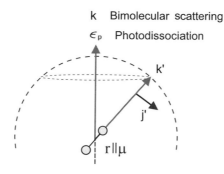

Axis of axial symmetry

Fig. 2.4. Vector diagram for a three-vector correlation between the axis of axial symmetry (the photon polarization vector ε_p and thus the molecular transition dipole moment μ for photodissociation, or the reactant relative velocity vector **k** for scattering) and product relative velocity vector **k'**. In this example μ is parallel to the bond axis **r**. In addition, the correlation between the polarization of **j'**, the product angular momentum, with **k'** and thus **k** is determined using a polarized laser probe of the product species.

(μ, **k'**), the Newton sphere surface pattern for photodissociation. Forward or backward reactive scattering is simply a (**k**, **k'**) surface pattern. For our bimolecular reaction AB + C $\rightarrow$ A + BC the group of before and after vectors (**k**, **k'**, **r**, **r'**, **j**, **j'**) and the two, three, and even the (thus far unobtainable) four-vector correlations between them have been described in detail by Case *et al.* [7].

Figure 2.4 illustrates a three-vector correlation measurable using imaging. The radius of the Newton sphere is the magnitude of the **k'** vector, while the surface pattern is the correlation with μ for photodissociation (or **k** for scattering). When using a polarized probe to detect the products making up the Newton sphere, we can also probe the directionality of a third vector in the process – for example **j'** in the (**k**, **k'**, **j'**) or (μ, **k**, **j'**) vector triads. The distribution of both **k'** and **j'** must have azimuthal symmetry about μ, but when a subset of **k'** vectors with a particular **j'** (or vice versa) is selected, this subset in general will not have azimuthal symmetry about μ. This limits the direct analysis of the image by inversion methods, as discussed later. The three-vector correlation quantifies these interrelations. Figures 2.5 and 2.6 show velocity map images in order to illustrate (**k**, **k'**, **j'**) and (μ, **k'**, **j'**) correlations.

In Fig. 2.5 the inelastic scattering process Ar + NO(j) $\rightarrow$ Ar + NO(j') reported by Chandler and coworkers [8] recently in *Science* is shown schematically. Vector **k** lies along the horizontal axis in the figure and **k'** defines the dashed circle. Using a very clever geometry the *sense of rotation* of the scattered NO product is measured as a function of **k'**. Molecules scattered to the right side of the Ar atom rotate preferentially counter-clockwise whereas those to the left rotate clockwise. In both

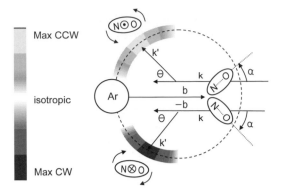

Fig. 2.5. Schematic centre-of-mass frame diagram and velocity map image of the inelastic scattering of Ar with NO(j) to form NO(j′), adapted from Chandler and coworkers [8] as published in Science. NO(j′) has a speed given by the dashed circle and is primarily forward scattered, in the direction of the NO(j) beam. The sense of rotation of NO(j′) is probed in the image, which is depicted as a slice through the scattering plane. The grey bar is coded for the degree of clockwise (CW) and counter-clockwise (CCW) rotation. As explained in Chandler *et al.* the specificity in rotation is induced by two sequential ('chattering') collisions of Ar with NO molecules oriented with angles of attacks given by α and impact parameter b in the figure.

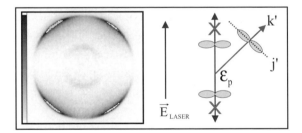

Fig. 2.6. Velocity map image of O(^{1}D) products from the photodissociation of molecular oxygen. Signal levels are indicated by the grey bar. The photodissociation and probe laser polarization ε_p lies in the vertical axis. The unusual angular distribution observed is due to a three-vector (ε_p, $\mathbf{k}'$, $\mathbf{j}'$) correlation, as explained in the text. The smaller inner disc is due to another process creating O$^+$ ions by O$_2^+$ photodissociation.

cases $\mathbf{j}'$ is perpendicular to $\mathbf{k}'$, but in addition, in this incredible experiment the *orientation* of $\mathbf{j}'$ is determined.

Another three-vector correlation is shown in Fig. 2.6 for photodissociation of molecular oxygen [9]. Here ϵ_p lies along the vertical axis of the figure and the larger disc in the image shows ($\mathbf{k}$, $\mathbf{k}'$, $\mathbf{j}'$) for O(^{1}D) atom products from the dissociation. These products are detected with a second laser polarized in the same direction as the dissociation laser. Because of strong angular momentum alignment of the atomic electronic angular momentum $\mathbf{j}'$ perpendicular to $\mathbf{k}'$, the vertically moving atoms cannot be efficiently ionized while atoms recoiling at larger angles θ are ionized. This leaves a hole at the poles of the $(\mathbf{k}, \mathbf{k}')$ angular distribution, which without the $(\mathbf{j}', \mathbf{k}')$ correlation would be $\cos^2\theta$. The $(\mathbf{j}', \mathbf{k}')$ correlation can be predicted by Wigner–Witmer rules [9].

2.2.8 Summary

A collision of a molecule with photons or another molecule creates Newton spheres. Reaction takes place and for a given set of possible internal states the two products fly in opposite directions on the surfaces of a set of nested Newton spheres. Polarization of the reactant angular momentum and velocity vectors creates anisotropic patterns on the surface of the Newton spheres. More information can be extracted from the surface pattern by probing the products with a polarization-sensitive detection process. Each reaction event creates only one set of points on the surface of paired spheres. By limiting position and velocity spreads in the reactant beams, and then repeating the scattering process many thousands of times a surface pattern emerges on each sphere that contains a wealth of information about the photodissociation, photoionization, or scattering process.

2.3 Images compared to time-of-flight methods

How are Newton spheres measured? There are many different ways and in this section imaging and a specific subset of time-of-flight (TOF) methods are compared. Both of these methods rely on creation and detection of ions but they often differ as to when the ions are created – at the formation point (imaging) or after the Newton spheres have expanded ('time-of-flight'). Table 2.2 lists several different types of experiments in molecular reaction dynamics for each method and the main characteristics of each experiment. These experiments and their defining parameters will be discussed in this section. We should point out that there are other powerful methods yielding information on Newton spheres that are not included in this table.

2.3.1 Detection of reaction products by ionization

Detection of products of uni- and bimolecular processes under single collision conditions is very difficult. We are limited to a handful of detection methods, each of which works best for a limited range of product types. These include surface ionization of alkali metal species; fluorescence detection following laser-induced fluorescence or chemiluminescence; ion detection following 'universal' electron impact ionization, pulsed laser ionization, threshold VUV–XUV ionization using synchrotron sources; and bolometric detection of the total energy of the product. Ionization methods have an important advantage over other methods in that the

Table 2.2. Selected characteristics of imaging and time-of-flight experiments

Method/Experiment	t_{event}	$t_{ionization}$	Mass selection	Velocity from:	Detection
Imaging					State Selective
Photodissociation	Dissociation laser	REMPI laser	Ion TOF	Image radius $\propto v \times$ ion TOF	Flux
Bimolecular	Random	REMPI laser	Ion TOF	Image radius $\propto v \times$ ion TOF	Density
Bimolecular – laser created reactant	Photolysis laser	REMPI laser	Ion TOF	Image radius $\propto v \times$ ion TOF	Density
Chemi-ionization	Random	t_{event}	other methods	Image radius $\propto KE^{1/2}$	Flux
Time-of-flight					All States
Photodissociation – electron impact detection	Dissociation laser	Distance of flight (DOF)/v	quadrupole	DOF/$v - t_{event} +$ ion TOF	Density
Bimolecular – electron impact detection	Random	DOF/v time-of-flight by $t_{chopper}$	quadrupole	chopper	Density
Bimolecular – laser triggered – electron impact detection	$t_{laser} +$ TOF reactant (w/vel spread)	DOF/v	quadrupole	DOF /$v +$ ion TOF	Density
Bimolecular – laser triggered + Rydberg tagging	Photolysis laser	tagging laser + field ionization	H** TOF	DOF/$v +$ ion TOF	Flux
Photodissociation – Rydberg tagging	Started by the dissociation laser	DOF/v	H** TOF	DOF/v	Flux

product charge/mass ratio, thus chemical identity, is determined. Fast and sensitive ion detectors are available and, by using electric fields, 100% ion collection efficiency is possible. Our selected TOF and imaging methods both use product detection by ionization. Neither method is superior; they differ in efficiency, resolution, applicability, etc.

2.3.2 All for one, or one for all?

Figures 2.7 and 2.8 illustrate the very different approach of TOF and imaging in capturing the Newton sphere. Crossed-beam scattering is chosen here as the example of the most challenging application for both techniques. For photodissociation the same diagrams apply if one of the reactant beams is a laser beam. In this case the centre-of-mass vector (see also Fig. 2.3) is parallel to the laboratory velocity vector of the molecular beam. The TOF method shown in Fig. 2.7 takes place after the neutral product molecules making up the Newton sphere have expanded and pass through a small entrance slit of the detector where ionization and mass selection

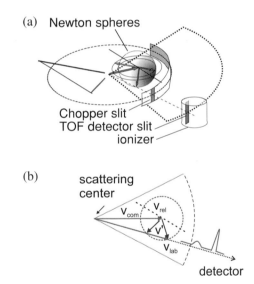

Fig. 2.7. (a) Time of flight (TOF) detection of Newton spheres from crossed beam scattering. The centre-of-mass and relative velocity vectors are shown as in Fig. 2.3. A detector with a vertical slit is rotated about the crossing point in the collision plane of two crossed molecular beams. Product molecules passing through a chopper slit and detector slit are ionized and mass-selectively detected. A typical forward scattering process is illustrated as a surface pattern on the sphere. (b) Vector diagram for detection of products in the lab frame. The angular position of the detector determines which laboratory velocities, $\mathbf{v}_{lab}$, are detected. $\mathbf{v}_{lab}$ is the vector sum of the centre of mass velocity $\mathbf{v}_{com}$ and the product relative velocity $\mathbf{v}'$ for the product Newton spheres. As shown, two positions on the Newton sphere pass to the detector at the selected laboratory scattering angle. Recovery of the Newton sphere information usually requires a complicated forward convolution of the TOF data.

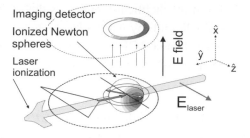

Fig. 2.8. Ion imaging detection of Newton spheres from crossed-beam scattering. Laser ionization of the product molecules takes place at the scattering centre and a static electric field is used to project the expanding sphere of ions (in the direction defined as the x axis) onto a two-dimensional imaging detector. The laser propagates along the y axis, chosen so that the z axis, the direction of electric field for linear polarization, lies parallel to the centre-of-mass velocity vector. See also the caption for Fig. 2.7.

takes place. Note that only a small solid angle of the Newton sphere is measured per angular position of the detector. Here *all* the effort is made for *one* scattering angle. The plane of scattering is probed, however, which contains all the necessary information as long as the angular momentum of the reactants is not polarized.

In imaging (Fig. 2.8) the product molecules are ionized *at the scattering centre* and then projected all at once onto the 2-D imaging detector. Mass selection is provided by the time-of-flight between ionization and detection, and the (detection) laser polarization is set parallel to the product relative velocity vector, the axis of cylindrical symmetry for scattering. Now, *one* measurement provides *all* the angle–velocity information.

Two advantages of imaging should be clear from comparing Figs. 2.7 and 2.8. First, the full Newton sphere of product molecules is probed and recorded, instead of a small angular fraction of a slice of the Newton sphere through the collision plane. Second, the centre-of-mass velocity results in a simple displacement of the Newton sphere in the imaging experiment, while it is convoluted with the Newton sphere information in the TOF technique. Other advantages and disadvantages of imaging versus TOF are discussed in the following subsections.

2.3.3 Event time zero

The instant of photoabsorption or collision is defined as t_{event}, event time zero. In photodissociation and photoionization studies t_{event} is simply the moment of crossing of a pulsed photolysis laser with the molecular beam, a time which is controlled within nanoseconds (to femtoseconds) by the experimenter. In studies of bimolecular scattering, event time zero can be any time that both reactant species are present in the scattering volume. For continuous molecular beams t_{event} is random in time. When using pulsed molecular beams the overlap time of the beams is still large compared to the time needed for mass selective detection. Pulsed lasers can be used to generate reactants by photodissociation. A time delay is needed to allow for the build up of products and the volume of photodissociation source for reactant formation can introduce a spread in the times when t_{event} occurs.

2.3.4 Mass selective detection and efficiency

The chemical identity of the product is always the first question in molecular dynamics. Mass selective ion detection is thus critical for both imaging and TOF methods. Electron impact provides 'universal' detection of products with an ionization efficiency of $\sim 10^{-4}$ but 'cracking' of higher mass species into the measured mass channel is often a problem with electron impact. In neutral time-of-flight detection a quadrupole mass filter after the ionizer is used to pass only ions of the correct charge-to-mass ratio. Quadrupole filters have a throughput of <40% of the selected ions.

In most imaging experiments the neutral-ion conversion is done with a pulsed laser which provides a well defined *time of ionization*, $t_{ioniz.}$, for use in ion TOF mass spectrometry. The ionization efficiency in this case can be as high as 100%. For mass selection, the imaging detector is turned on when the desired mass arrives, as discussed in Section 2.4.5.6. The throughput of ion TOF detection using velocity mapping is 100%. These high efficiencies can be misleading, however, since the important factor is the signal/noise ratio. 'Noise', e.g., due to non-selective ionization and background gas problems, often increases much faster than signal at high ionization laser powers.

2.3.5 Determination of product internal energy and velocity

In the time-of-flight method illustrated in Fig. 2.7 ionization takes place at the detector after the neutral Newton spheres have fully expanded. In the laboratory frame the products scatter at an angle defined by the sum of the centre-of-mass and relative velocity vectors. For a range of laboratory detection angles two parts of the Newton sphere can enter the detector, as illustrated in Fig. 2.7b. Extraction of centre-of-mass frame information requires indirect forward convolution analysis. In order to determine the product internal energy distribution a chopper wheel is placed in the path between the scattering centre and the detector. This provides a sharp time zero ($t = 0$, small Δt) for the velocity measurement. Particles with higher internal energy will have less kinetic energy and thus arrive at the detector later in time, thus at a longer time-of-flight.

In ion imaging (Fig. 2.8) the neutral products are first state-selectively ionized and then projected by an electric field (x direction in the figure) onto the imaging detector. The distance-of-flight (DOF) of the ion perpendicular to the projection axis (y, z direction) is recorded after a fixed time Δt (the arrival time at the detector of the desired ion species). A product state with higher internal energy will have less kinetic energy and thus cover less distance from the middle of the detector.

Using TOF, the ion signal $S(t)$ is recorded as a function of time after the neutral product has moved over a fixed distance Δd. With imaging, the signal is recorded as a function of radial distance covered $S(d)$ after a fixed flight time Δt. $S(d)$ is a limiting resolution factor in imaging due to the detector characteristics, as discussed later. Velocity is obtained directly in ion imaging ($v = S(d)/\Delta t$) while in time-of-flight the inverse velocity is measured ($1/v = S(t)/\Delta d$).

2.3.6 Flux versus density detection

The desired *rate* of product formation in a molecular dynamics process is derived from a product *flux* measurement. In conventional use, pulsed laser detection methods such as laser ionization and laser-induced fluorescence yield the product *density* instead of flux. Liu and coworkers [10] give a detailed discussion of the effects of the product laboratory velocity when using laser detection methods. They point out that if all product molecules can be monitored regardless of their laboratory velocity, a flux mode of collection can be realized. For the following discussion on flux versus density detection consider Fig. 2.9, where product molecules are probed by an ionization laser probe beam focused to a 100 μm diameter. This geometry is

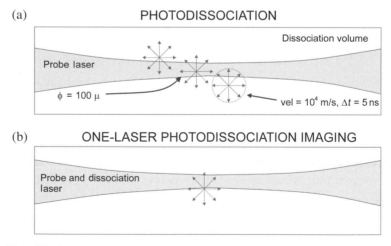

Fig. 2.9. Simplified geometry for detection of photodissociation products using (a) separate pump and probe lasers and (b) a single laser for both dissociation and ionization. Products are formed in a larger region and detected within the focus region, which has a minimum diameter of 100 μm. The shown Newton spheres diameters correspond to molecules moving with speeds of 10^4 m/s in 5 ns.

typical when using $(2 + 1)$ resonance enhanced three-photon ionization. With one-photon dissociation using a larger diameter dissociation laser beam, and crossed-beam bimolecular scattering, this probe laser diameter is much smaller than the diameter of the production region. For a scale perspective, in 5 ns products with a velocity of 10^4 m/s (fast!) yield Newton spheres of the diameter shown in Fig. 2.9.

2.3.6.1 Production by laser dissociation

If we ignore the finite pulselength of the photodissociation laser, all Newton spheres are the same size at the instant of ionization by the probe laser. With a 5 ns delay between dissociation and ionization the Newton sphere of our fast products is larger than the probe diameter. Still, as many fragments have entered the probe volume from outside as those that have left it, so all parts of the Newton sphere are equally probed. The Doppler shift must also be accounted for, as described in Section 2.4.4.3. This is a flux detection mode.

2.3.6.2 Production and probing in a single-laser experiment

In the simplest type of photodissociation imaging experiment, illustrated in Fig. 2.9b, the same laser used for state-selective ionization detection also drives a photodissociation step, creating the neutral species that the ionization laser is set for resonant detection. This sort of one-laser signal creates background signal in two-laser experiments and thus needs extra attention. Dissociation and ionization must both occur within the ~ 5 ns laser pulselength. Particles moving along the molecular beam axis stay in the ionization beam and have a high chance of detection while energetic $(>10^4$ m/s$)$ neutral photofragments traveling perpendicular to the laser beam axis can completely escape the ionization step. In this case a portion of the Newton sphere is not detected at all, thus neither a flux nor a density mode is realized. This effect is illustrated in more detail in Section 2.4 and in Fig. 2.17.

2.3.6.3 Production by bimolecular scattering

We have already mentioned that the creation of products by bimolecular scattering takes place on a long time-scale compared to photodissociation due to the length of the molecular beam pulses. As can be imagined from Figure 2.7b it is possible to create zero-velocity products in the lab frame $(\mathbf{v}_{lab} = 0)$ when the product relative velocity is equal and opposite to the centre-of-mass velocity. If the time period for scattering is long compared to the detection time, low velocity products will build up in the detection zone and be more easily detected than products with high laboratory velocity. Under ideal steady-state formation conditions, the chance that a molecule lies within the probe laser beam at the instant of probing is inversely proportional to laboratory velocity. This is the (normal) density detection mode using pulsed lasers. A simple example of laboratory velocity effects is illustrated in

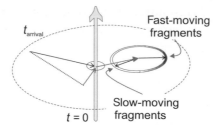

Fig. 2.10. Imaging detection of products from crossed-beam scattering by a laser propagating perpendicular to the scattering plane. The laser polarization is set along the relative velocity vector and the beam is intercepted before striking the imaging detector. Products with low laboratory velocity are more efficiently detected.

Fig. 2.10 for detection of scattering products using a laser propagating perpendicular to the scattering plane. Because products with low scattering laboratory velocity remain longer in the detection volume, they are detected more efficiently than those with high velocity. The low velocity species will thus have a higher measured concentration than those with high velocity, and a complicated correction function is necessary. Such a function has been described and applied to the inelastic scattering data of Chandler and coworkers [8] shown schematically in Fig. 2.5. In their study the laser detection geometry was in-plane, as drawn in Fig. 2.8. The density–flux transformation can be even more complicated in this geometry because fast products moving perpendicular to the ionization laser direction can escape detection. Fast species are less likely for this endoergic inelastic scattering process.

2.3.7 Time of ionization and types of ionizers

At present, three different ionization schemes are combined with neutral time-of-flight methods. Electron impact, the original method, is applicable to many species and is mass selective over a wide mass range when coupled with a quadrupole filter. The latest generation electron impact detectors using liquid helium cooling around the reaction volume. These detectors have extremely low background noise. 'Cracking' is the biggest problem with electron impact – daughter ions appear along with or instead of the parent ion, and their yield can be a complex function of the internal energy of the product species. Because cracking is dependent on the product internal energy state, electron impact ionization often deviates from the ideal case of being state-independent. With electron impact the time spread of ionization, $\Delta t_{\text{ionization}}$, can be broad, as determined by v'/L with v' the neutral product velocity and L the length of the ionizer.

Lately, synchrotron light sources producing narrow-band tunable XUV light in the 10–15 eV energy range provide an alternative to electron impact ionization.

With these user-facility sources ultra-sensitive detection of species with low ionization potentials is possible by tuning the photon energy below that of the major background species. State-selective ionization is not likely for most molecules, but reasonably state-independent ionization can be expected for almost any type of species. Again, quadrupole filters provide the mass selectivity, and, like electron impact, $t_{ionization}$ is (for most synchrotron sources) not sharply defined. Both XUV photoionization and electron impact ionization provide density detection.

For detection of H atom products a powerful ionization method is available that applies field ionization of highly excited Rydberg states. Unlike the previous two approaches for neutral time-of-flight, the H atoms are excited ('tagged') in the product source volume to a high-lying Rydberg state by VUV–UV lasers. This allows excitation of all H atoms simultaneously. Thus far H-atom tagging has been combined only with TOF detection at discrete angles using field ionization followed by a charged particle multiplier. Speed resolution using this approach is superior ($\Delta v/v < 0.1\%$) to all other imaging and TOF methods since t_{event} is defined by a nanosecond laser (small $\Delta t/t$), and the Rydberg atoms can fly a great distance (with a small $\Delta x/x$) before detection (which is essentially $t_{ionization}$). Because the field ionization efficiency can be made independent of the H atom velocity, the method is a flux detector. The detection process is limited at present to H atoms. However, see the papers by Wodtke and co-workers [11] in which CO from ketene photolysis was detected by excitation to a metastable triplet state.

Laser ionization is used in conjunction with the ion imaging method. In most cases REMPI (Resonance Enhanced Multiphoton Ionization: $A + nh\nu \rightarrow A^+ + e^- + TKER$, with $n \geq 2$) is used for state-selective ionization. Direct one-photon ionization using a VUV F_2 excimer laser (157 nm) is also possible for non-state-selective detection of species with IP < 7.8 eV. REMPI is by no means a universal detector. The important OH radical, for example, is extremely difficult to detect using REMPI. REMPI also has a small duty factor due to the low repetition rate (<100 Hz) of pulsed lasers and is thus best combined with pulsed molecular beams. With laser photolysis and ionization the experimenter has control of both t_{event} and $t_{ionization}$, however, which allows both efficient and informative collection of the Newton spheres.

In some experiments, such as Penning and associative ionization, the products detect themselves. All ions are collected regardless of their velocity, thus with a flux detection mode. In our velocity map image shown in Fig. 2.3, products are formed as ions during the time of overlap of the metastable Ar* and Ne beams. In this case $t_{event} = t_{ionization}$ and t_{event} is not under experimental control. Mass selectivity by ion TOF is not possible in this case thus both product masses (Ar^+ and $NeAr^+$) appear at the detector at a position determined by the $KE^{1/2}$ of the species.

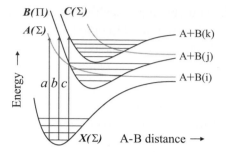

Fig. 2.11. Hypothetical potential energy curves for different types of photodissociation pathways. (a) Direct excitation from a $X(\Sigma)$ ground state to a repulsive state $A(\Sigma)$ correlating to the first dissociation limit A + B(i). (b) Excitation to the repulsive wall of state $B(\Pi)$, a bound state correlating with the second dissociation limit A + B(j). Curve crossing to the first dissociation limit via the repulsive state excited in pathway (a) is possible. (c) Excitation at an energy exceeding the first and second dissociation limits to a bound state $C(\Sigma)$, correlating to A + B(k). Predissociation can take place by curve crossing via another repulsive state to the second limit as drawn in the figure.

2.4 Velocity map imaging of photodissociation

Thus far, the greatest impact of velocity map imaging in molecular dynamics research has been in the field of photodissociation of diatomic and small polyatomic molecules. Small molecules dissociate into smaller fragments with a limited number of internal energy states, leading to widely spaced and thus more resolvable product Newton spheres.

Potential energy curves for photodissociation of a hypothetical diatomic molecule AB into fragments A and B (internal state i, j, or k) shown in Fig. 2.11 are excited via three different pathways a, b, and c. Such curves are constructed for each molecule using molecular quantum theory and models such as the Wigner–Witmer rules [12]. Due to spin-orbit coupling and other effects, curve crossing can take place leading to other dissociation products than those predicted by the molecule – separated atom correlation diagrams. Using this diagram as a guide, we can make a 'wish list' of the most important photodissociation properties we wish to quantify, as a function of excitation wavelength:

- D_0, the A–B **bond energy** of the X state, and of the B and C bound excited states such as those reached by pathways b and c.
- The **symmetry** (Σ, Π, Δ, ...) of the ground and excited states.
- The **absorption cross sections** σ_a, σ_b, σ_c for each transition.
- Time-scales for **(pre-)dissociation** of each quantum level of the excited states.
- **Product yields** to the first, second, and third dissociation channels.
- **Angular distributions** (beta parameter, β, and higher parameters) and possible **vector correlations** for each dissociation channel.

- Characterization of the importance of **curve crossing** from the photoexcited curve to curves leading to other dissociation products.
- Characterization of the importance of **coherent** ($a + b + c$) versus **incoherent** (a, b, or c) **excitation** of the different pathways.

Each of the above-listed properties is a research field in its own right. All provide important tests of molecular quantum mechanical theory, and many have practical applications, for example, in atmospheric chemistry. Since the emphasis of this chapter is on the experimental method, only a few of these parameters will be discussed in any detail.

All of the above information is extracted from product photofragment images in the form of radial, angular, and three-vector information. From the product KER and from energy balance the internal state distribution of both fragments can be determined, as illustrated in Fig. 2.2. The surface pattern of the Newton sphere of photodissociation fragments is characterized (Table 2.1) via the beta (β) parameter and higher order alignment parameters as needed. From these two main quantities, KER and β, and from knowledge of the spectroscopy and quantum mechanics of the parent molecule and fragments we begin to fill in as much as possible of the above 'wish' list.

2.4.1 The beta parameter

Direct dissociation of a diatomic molecule occurs on a time-scale much shorter than a classical rotation period. The fragments fly off parallel or perpendicular to the polarization direction of the light field with an angular distribution given by the equation we have already met in Chapter 1:

$$I(\theta) = (1 + \beta P_2(\cos\theta))/4\pi. \tag{2.1}$$

Here θ is the angle between the fragment velocity vector and light polarization direction as shown in Fig. 2.1, and $P_2(x) = (3x^2 - 1)/2$ is the second-order Legendre polynomial. [For multiphoton transitions, higher order Legendre polynomials are needed, in general $I(\theta) \propto \Sigma_n a_{2n} P_{2n}(\cos\theta)$ and thus the anisotropy parameters $a_2 = \beta$ and $a_4 = \gamma$ for a two-photon process.] The factor 4π normalizes the function for the full solid angle. β in this expression is equal to 2 for a parallel transition and -1 for a perpendicular transition. Filling this out in equation (2.1) leads to $I_{//}(\theta) = (3/4\pi)\cos^2\theta$ and $I_\perp(\theta) = (3/8\pi)\sin^2\theta$, as shown before in the example of Newton spheres with polar and equatorial surface patterns. In general, the value of β lies in between these extreme values, and the angular distribution can be written as a simple linear combination of parallel and perpendicular: $I(\theta) = A\cos^2\theta + B\sin^2\theta$, where $A = (1 + \beta)/4\pi$ and $B = (1 - \frac{1}{2}\beta)/4\pi$. For $\beta = 0$, the angular distribution is isotropic in space.

In the case of mixed absorption such as the overlap of pathways a, b, and c in Fig. 2.11, β lies between the two extremes. By measuring the angular distribution of the fragments, the percentage of parallel versus perpendicular character can be determined. The beta parameter is then a linear combination of the contributions from possible reaction channels. In the example sketched in Fig. 2.11, let us assume that pathway a is a pure parallel transition to state $A(\Sigma)$ with absorption cross-section σ_a and b is a pure perpendicular transition to state $B(\Pi)$ with absorption cross-section σ_b. The probability of curve crossing from state $B(\Pi)$ to $A(\Sigma)$ at their intersection is P_{BA}. The contributions to the first dissociation limit lead to $\beta = \sigma_a\beta_{//} + \sigma_b P_{BA}\beta_\perp = 2\sigma_a - \sigma_b P_{BA}$. Measurement of the β parameter is therefore a measure for absorption cross-sections and curve crossing probabilities. These can be unraveled from the combined measurement of the anisotropy of and the branching between all dissociation limits B(i, j, and k).

Predissociation, as in pathway c to limit B(j) or pathway b to limit B(i) in Fig. 2.11 is not instantaneous. Should the molecule rotate before dissociation the degree of correlation between the electric field vector of the light and the fragment recoil direction will be lowered, but not completely to zero. For predissociation when the lifetime τ of the excited state is similar to the parent molecule rotational period T and the according rotational frequency $\nu = 1/T$ of that state, the anisotropy parameter is given by [13]:

$$\beta_{//}(\tau) = 2(\nu^2\tau^2 + 1)/(4\nu^2\tau^2 + 1) \qquad (2.2)$$

for a parallel transition, and

$$\beta_\perp(\tau) = -(\nu^2\tau^2 + 1)/(4\nu^2\tau^2 + 1) \qquad (2.3)$$

for a perpendicular transition. The limiting values for these anisotropy parameters are $\beta_{//} = 0.5$ and $\beta_\perp = -0.25$. Measurement of β for each final product channel thus provides information on the symmetry of the electronic states involved in the absorption process for direct dissociation, and the lifetime of a single state with respect to rotation for predissociation.

In the simplest picture of photodissociation, optical excitation selects molecules in the lab frame via their directionality of μ and, because of the $\mu - k'$ correlation, the direction of the dissociation product velocity. Should curve crossing take place the relative yields of the products will change but their angular distribution is determined by the symmetry of the optically excited state. Using this simple model it is possible to decompose overlapped absorption spectrum continua for most simple molecules.

As mentioned before, multiphoton excitation, product vector correlations, and pre-alignment of the parent molecule before photodissociation may create angular distributions that cannot be fit by only the beta parameter. A general description of data analysis in this case is given in Chapter 4 of this book. Furthermore, pathways a, b, and c can be coherently excited, especially when the corresponding absorption cross-sections σ_{abc} are similar in magnitude. Coherent excitation leads to a rich and complex outcome in correlated product helicity and other effects, as described in Chapter 4, and by Zare and coworkers [14].

2.4.2 The velocity map imaging experiment

Imaging involves the experimental steps outlined in Fig. 2.12: (a) creation of the spheres by photodissociation; (b) conversion of the photofragments to ions by laser ionization; (c) projection of the ionic Newton spheres onto a 2-D detector; and (d) recovery of the three-dimensional information from the 2-D image using a mathematical transformation described in detail in the next chapter. The result of (d) should be equivalent to taking a slice through the raw Newton sphere formed in step (a). Lately, considerable effort is being spent on 'slicing' the experimental 3-D ion distribution, yielding (d) directly from (b), and which is described in more detail in the final chapter of this book.

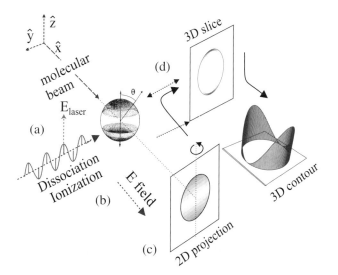

Fig. 2.12. Schematic diagram of the photofragment imaging approach to measuring Newton spheres from photodissociation. (a) Photodissociation of molecules in a molecular beam by using a linearly polarized laser with the polarization direction parallel to the detector face. (b) Conversion of the photofragment molecules making up the Newton spheres into ions by laser ionization. (c) Projection of the ion spheres onto a two-dimensional (2-D) detector. (d) Mathematical transformation of the 2-D image back to the three-dimensional data of step (a). This slice through the middle of the Newton sphere is displayed in (d) as either a false color 2-D diagram or 3-D contour diagram. With experimental slicing techniques discussed in Section 2.4.6.1, it is possible to avoid step (c).

and probe induce parallel transitions, the probe recoil will add and subtract energy along the vertical direction (where most of the products are) and will have little effect for the few products moving in the perpendicular direction. For case (b) the situation is similar except both dissociation and probe induce perpendicular recoil. Cases (a) and (b) will lead to a splitting of the nascent velocity into two components. The mixed cases (c) and (d) will have less effect along the primary direction of the nascent velocity but will tend to smear out the slice for larger angles away from the poles [case (a)] or equator [case (b)]. This will affect the measured product angular distributions.

We have found experimentally that the photoelectron in the $(2 + 1)$ REMPI of $O(^3P_2)$ atoms at 226 nm is ejected in a perpendicular or $\sin^2\theta$ distribution, where θ is the angle between the recoil velocity vector and the polarization direction of the linearly polarized REMPI laser. The REMPI excited electron is in the $3p$ manifold and interference of the outgoing s and d waves yields a perpendicular angular distribution, just as in ionization (photodetachment) of a ground state $2p$ electron in O^-. Recoil due to the REMPI probe step will thus be perpendicular to the laser polarization axis while the nascent velocity distribution is that of a parallel transition, which is shown in the bottom quadrant of Fig. 2.14.

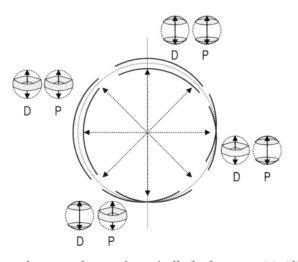

Fig. 2.14. Newton spheres are shown schematically for four cases (a)–(d) for perpendicular and parallel dissociation/ionization processes. Shaded areas on the spheres indicate regions (directions) of maximum product. A slice through a Newton sphere with a vertical symmetry axis is shown in the centre of the figure as a dashed circle. This ring will be split due to the extra effect of recoil induced in the probe process. In the top quadrant a parallel dissociation (D) is perturbed by a parallel probe (P) recoil. In this case the products fly primarily along the vertical axis and are thus split by the recoil process. The right, bottom and left quadrants show the effects of perpendicular D, parallel D, perpendicular P, parallel P; and perpendicular D and P recoil, respectively.

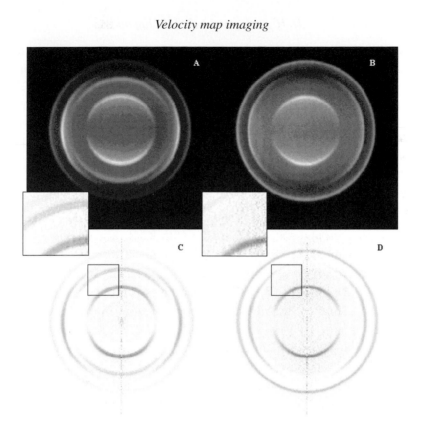

Fig. 2.15. H-atom photofragment image following excitation at 277 nm and ionization at 243.17 nm at vertical (A,C) and horizontal (B,D) polarization of the probe laser. The lower panels (C and D) show the inverse Abel transform of the raw image shown above. The splitting of the H-atom signal from the parallel H+I* channel only appears when the polarization of the probe laser is perpendicular to the TOF axis (see insets).

Another example of REMPI recoil effects can be observed in the photoionization of H atoms from photodissociation of HI in the near-UV region [18]. Raw and inverted images of H^+ detected by 243 nm $(2 + 1)$ REMPI of H atoms produced from the photodissociation of HI^{17} at 277 nm are shown in Fig. 2.15. Since the probe laser at 243 nm also dissociates HI, a second set of I and I* signals with higher kinetic energy release is observed together with the pump and probe signal. Note the splitting of the contribution from the parallel dissociation channels (the two 'polar' rings) for both the 243 nm and 277 nm signals. The velocity splitting is equal for both channels and corresponds to 425 m/s, matching the observed ~ 3 pixel splitting of each polar ring.

REMPI recoil follows a $\cos^2\theta$ distribution, because in the final ionization step a $2s$ electron is ejected into the p continuum. For the parallel dissociation channels (Fig. 2.14, top quadrant) this results in the most easily observed probe laser induced recoil splitting of the Newton ring.

The splitting is not observed when using a REMPI probe laser with the polarization set perpendicular to the detector face. In this geometry the extra recoil is projected along and in the opposite direction to the time-of-flight axis. Ring splitting by the probe step is not seen because the immersion lens crushes the image in this direction. The image resolution is, however, lowered by the probe recoil effect. As shown above, such recoil effects must be considered for all cases of REMPI.

2.4.4.3 The Doppler shift

Doppler shifts occur only for bound–bound transitions of recoiling product species. As an example, a ground state $O(^3P_2)$ atom with KER (kinetic energy release) of 1.95 eV recoils with a velocity $v = 4850$ m/s. For the resonant $(2p \rightarrow 3p)$ part of the $(2 + 1)$ ionization, this introduces a Doppler shift of $(1 \pm v/c)^{-1} hc\nu = 1.43$ cm^{-1}. The $2p\,^2P_{3/2} \rightarrow 3p\,^2P_{3/2}$ transition is centred at 88 630.8 cm^{-1} for atoms recoiling along the REMPI laser beam direction. In order to detect all the atoms with equal efficiency it is thus necessary to scan the laser over a wavelength range of at least $88\,629.4 - 88\,632.2$ cm^{-1}.

Consider a simple one-laser photodissociation imaging experiment where molecular oxygen is photodissociated by 225 nm light tuned to the $2p \rightarrow 3p$ transition. Photodissociation of O_2 at this wavelength is a bound-free transition into the Herzberg continuum, which is extremely weak and essentially wavelength-independent in the $88\,629.4-88\,632.2$ cm^{-1} range. Two $O(^3P)$ atoms are created in the process; these are immediately ionized due to the REMPI process. The dissociation and ionization lasers are by definition perfectly overlapped. Also observed in this process are two larger rings corresponding to $O(^3P_2)$ atoms created by *two-photon* dissociation to the $O(^3P) + O(^1D)$ and $O(^3P) + O(^3P)$ limits, respectively. These atoms recoil with 1.95 and 2.93 eV KER. Using velocity map imaging the ions formed in one laser pulse are projected onto a double MCP detector which converts one O^+ ion into 10^7 electrons. The electrons are accelerated towards a phosphor screen where they create a localized visible flash of light that is recorded by a CCD camera. The output of the camera (a single shot O^+ image) as a function of the wavelength of the photodissociation/REMPI laser is shown in Fig. 2.16. By summing images from several laser shots and then repeatedly stepping the laser wavelength well above and below the Doppler-free centre wavelength, a total image is obtained, as shown on the right side of the figure.

2.4.4.4 Flux-density transformation

Laser ionization does not always provide the desired product flux detection in photodissociation imaging experiments. The dangers of a one-laser photodissociation/photoionization experiment have been discussed using Fig. 2.9b. The same

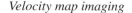

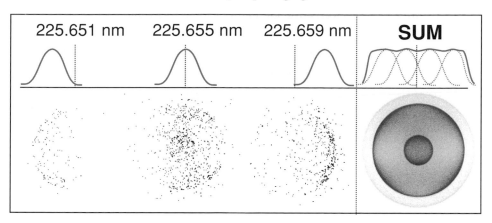

Fig. 2.16. Illustration of the Doppler effect in detection of $O(^3P_2)$ atoms by $(2 + 1)$ REMPI at 225.655 nm. The laser bandwidth is represented by a Gaussian-shaped curve and the ions formed from 10 shots of the laser at different wavelengths are shown in event counting mode. The summation of 10 000 laser shots is shown on the right side of the figure. The inside ring of the figure corresponds to one-photon dissociation while the outside (faster) ring is from two-photon dissociation to the $O(^3P) + O(^1D)$ limit. A weaker ring corresponding to two-photon dissociation to the $O(^3P) + O(^3P)$ limit is seen at the largest radius.

effect arises in a two-laser photodissociation experiment when the time delay between photodissociation and photoionization is longer than r_{probe}/v where r_{probe} is the radius of the ionization laser beam and v is the product speed. Consider a Cl atom photofragment with 1.13 eV KER, thus a velocity of 2500 m/s or $\sim 100 \ \mu m$ in 40 ns. With dissociation–probe time delays greater than 40 ns we should begin to lose those molecules travelling perpendicular to the detection laser axis, and at longer time delays only those travelling exactly along the laser axis will be seen. Figure 2.17 shows this effect for images of Cl from Cl_2 photodissociation around 300 nm and photoionization of Cl atoms using $(2 + 1)$ REMPI at 235 nm.

2.4.5 Conversion of ions to images

2.4.5.1 Detection geometry

With the above precautions we now have our expanding sphere of ions. How do we capture these spheres' mass selectively and with minimal loss of information? For imaging there are two sometimes conflicting demands on the apparatus; first the ion packets for different masses should be well separated, and second, the ion KER should be accurately mapped onto the position sensitive detector. Furthermore, the ion packet should either arrive at essentially the same time (time focusing or 'pancaking'), otherwise, the ion sphere packet should arrive sufficiently and uniformly spread out in time so that the full 3-D information can be resolved using

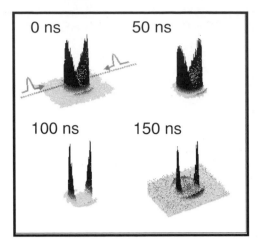

Fig. 2.17. Raw velocity map images for Cl^+ atoms from Cl_2 photodissociation as a function of time delay between photodissociation and photoionization. For time delays greater than 50 ns the angular distribution begins to peak strongly along the laser propagation direction. A weaker Cl^+ signal independent of the time delay arises from the photodissociation laser only.

time-resolved gating of the channel plates. Time gating of the detector is discussed in a following section.

For optimal velocity mapping resolution the molecular beam is pointed directly onto the imaging detector, as shown schematically in Fig. 2.12 and in more detail in Fig. 2.13. This geometry takes advantage of the narrow transverse velocity distribution of the pulsed and skimmed molecular beam ($T \sim 1K$) compared to the longitudinal velocity spread characterized by the speed ratio ($\Delta v / v \sim 10$). The laser beam crosses the molecular beam at right angles, forming an ionization volume as wide as the molecular beam diameter. For particle correlation experiments the need for dual detectors precludes the high-resolution on-axis geometry. In these state-of-the-art imaging experiments pioneered in the laboratory of Carl Hayden at the Sandia Livermore National Laboratory, the position and arrival time of the mass-selected photoion is measured in relation to its partner photoelectron in femtosecond pump–probe experiments.

2.4.5.2 Ion optics

Electrostatic lenses are used to direct the expanding ion spheres onto the position sensitive imaging detector. Figure 2.18 shows several different electrostatic lens configurations used in imaging experiments. Configuration (a) is that of Wiley and McLaren [19], (b) is a simpler imaging electrode version used for ion imaging and discussed in reference [19]; (c) is that of velocity map imaging, and (d) is an Einzel lens configuration that can be used in conjunction with velocity map imaging.

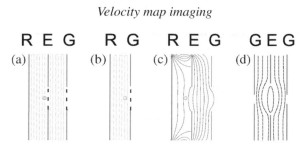

Fig. 2.18. Cross-section and field lines for electrostatic lenses used for ion imaging. (a) The original ion imaging configuration using fine mesh covering the flat annular Repeller (R), Extractor (E), and Ground (G) electrodes. The laser crossing volume is shown as a shaded circle. (a) Ion imaging, (b) one-grid imaging, (c) velocity map imaging, (d) Einzel lens for magnification of velocity mapped images.

The Wiley–McLaren technique (Fig. 2.18 a) uses two flat annular ring electrodes covered with a very fine mesh grid to form a homogeneous electric field along the TOF axis. This configuration is optimized to ensure that all ions of the same mass arrive at the detector at the same time regardless of their point of formation in the ionization region (space focusing). For the simple one grid imaging setup of Fig. 2.18b the space focusing condition is simply $D = 2s_0$, which leads to an impractical short value for D. For the two-grid configuration of Fig. 2.18a with 15 mm lens spacing and the dimensions labelled in Fig. 2.13, typical voltages on the repeller (E_R), extractor (E_E), and ground (E_G) electrode are 5000 V, 3000 V, 0 V, respectively, where the E/R ratio is chosen to match the Wiley–McLaren space focusing condition

$$D = 2s_0 k_0^{3/2}\left(1 - d/s_0(k_0 + k_0)\right)^{1/2}; \quad k_0 = (s_0 E_R + d E_E)/s_0 E_R. \quad (2.4)$$

For the dimensions $s_0 = 7.5\,\text{mm}, d = 15\,\text{mm}, D = 360\,\text{mm}$, the required $E_E/E_R = 0.66$. This space focusing condition, however, does not lead to velocity mapping, which is described later in this section. It should also be pointed out that for Wiley–McLaren conditions the ionization volume should be located extremely close ($< 2\,\text{mm}$) to the extractor plate.

For a single photofragment recoil KER resulting in a velocity, v_y, perpendicular to the detection axis, most of the ions lie at the outer edge of the disk forming a circle of radius

$$R \approx v_y \times \text{TOF} \approx L(\text{KER}/qV)^{1/2} \quad (2.5)$$

where L is the time-of-flight tube length, q the particle charge, and V the electrostatic potential with respect to the imaging detector. The arrival time of the particle is given by

$$t \approx L/v_x \approx L(m/2qV)^{1/2} \quad (2.6)$$

where m is the particle mass. Equations 2.5 and 2.6 are correct for all charged particles, regardless of their mass. In other words, with the detector time-gated at the TOF t for the proper mass, any species with the same kinetic energy will appear at the same radius R on the detector. This is quite helpful in calibrating signals where a molecular parameter, the bond dissociation energy, for example, is unknown. Note also that $KE^{1/2}$ is mapped instead of momentum or velocity.

The grids are removed from the electrodes in velocity map imaging (Fig. 2.18c). Velocity map imaging has the same ion acceleration properties as ion imaging except that a magnification factor, M, appears in the equation for R. For most apparatus $0.9 < M < 1.4$. This magnification factor, and the proper repeller/extractor E_E/E_R ratio for a set time-of-flight length, depends on the apparatus and is easily found by experiment or by trajectory simulation. A trajectory simulation for our apparatus with $L = 36$ cm, $V_R = 1000$ V, $V_E = 750$ V is shown in Fig. 2.19.

As seen in the lower panels of the figure, trajectories corresponding to initial velocity $\{|v_x|, v_y, v_z\}$ arrive at the same point on the lens focal plane, regardless of their point of formation within the simulated 3 mm spread in origin along the laser (z) axis. This is the magic of velocity map imaging – the ability to map velocity independent of origin of formation, and without grids. Transmission is thus 100% and without grid distortions. Figure 2.20 illustrates, using raw images, the

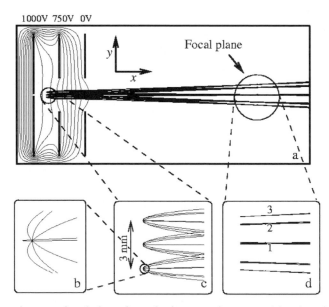

Fig. 2.19. Ion trajectory simulations for velocity mapping setup (a) using the 'Simion' ion trajectory program. Three different starting positions spaced 1 mm apart (c) simulate the laser–molecular beam crossing volume. Trajectories at every 45 degrees in the x–y plane begin quickly to bend in the field direction (b) and are focused at the velocity mapping focal plane (d), where the detector is placed.

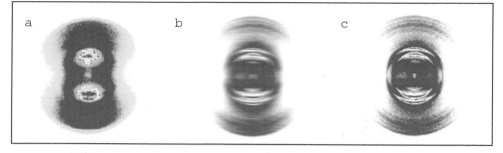

Fig. 2.20. Comparison of ion imaging and velocity mapping images taken under the same conditions. Panel (a) is ion imaging, panel (c) velocity map imaging, and panel (b) shows the effect of cancelling the velocity mapping by smearing the image out along the x direction by 2 mm. This illustrates the extra effect of the grids on lowering the image quality.

practical improvement in ion imaging (Fig. 2.20a) resolution provided by velocity map imaging (Fig. 2.20c) for identical conditions. A factor of ~ 20 better resolution is obtained. This is due to both the mapping and grid-free conditions. Figure 2.20b shows an image resulting from 'smearing' Fig. 2.20c along the x direction by 2 mm, which is the length of the laser beam/molecular beam crossing. The resolution under these de-mapped conditions is still superior to ion imaging, which emphasizes the deleterious effects of grids.

A typical Einzel lens configuration is shown in Fig. 2.18d. Equipotential (grounded) gridless electrodes sandwich another annular ring with a larger opening, set at voltage E_E. By placing the Einzel lens after the velocity map imaging lenses a magnification of the image by a factor of up to 20 is possible, as shown by Vrakking and coworkers [20]. An Einzel lens has a positive focus, however, meaning that the ion packet will be focused through a point on the TOF axis after the Einzel lens before reaching the detector. This can lead to space charge problems.

2.4.5.3 *Space charge*

Even when the electrostatic lenses are set correctly for velocity mapping, many factors can still degrade the observed ion image. Space charge, the repulsion of ions by other ions due to a too-high ion density, is always the greatest problem, especially since the detected ions can be perturbed by ions of a different, undetected, mass. Using a focused laser beam a rule of thumb is that the formation of more than ~ 100 ions (of any charge/mass ratio) leads to space charge. Slow-moving ions are of course most affected by space charge. Space charge in the formation volume leads to expansion of the image along the dimension of highest density, thus an oval-shaped image. Space charge is even possible at the imaging detector if many ions focus to the same fragment velocity. This leads to a broadening of the ring and lowering of the image resolution.

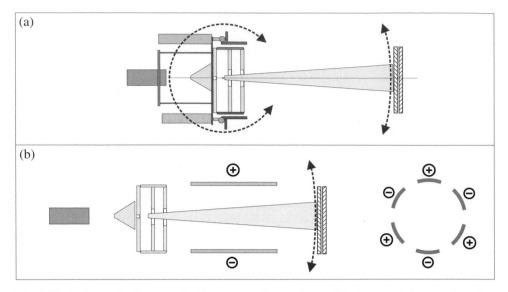

Fig. 2.21. Schematic diagrams for fine-tuning the position of the ion packet on the imaging detector. (a) Mechanical pivoting of the pulsed valve/skimmer/ion optics assembly about the laser–molecular beam crossing point using motorized micrometers, (b) a hexapole electrostatic deflection tube for deflection by a homogeneous field at arbitrary deflection angle.

2.4.5.4 Steering

Even with the best of apparatus design it is not so easy to hit the middle of the detector with the ion packet. Two different methods in use to actively steer the ion packet are shown in Fig. 2.21. Method (a), in use in our apparatus, pivots the whole pulsed valve/skimmer/electrostatic lens assembly in two dimensions about the laser–molecular beam crossing point. Motorized or mechanically controlled micrometers fine-tune the positioning with high accuracy. In method (b), described by Suzuki and coworkers [21], electric fields in a hexapole configuration are adjusted to provide a uniform electric field for any chosen deflection angle. The hexapole device should be of large diameter and used in a pulsed mode to avoid end-effects when the ion packet leaves the deflector.

2.4.5.5 Imaging detectors

The ion packet has now arrived at the front surface of the detector in good condition. We want to record the exact (y,z) position of each ion in the ion packet. An arrival time between 0.5 and 20 μs after laser ionization is typical for most combinations of charge/mass ratio, lens voltage, and flight tube length. For standard velocity map imaging acceleration by the electrostatic lens has 'pancaked' the ion sphere into a (y,z) disc ~ 20 ns wide in time along the detection (x) axis. Because the laser has probably created ions at other masses, we want to turn the detector on for a short period in time in order to detect one selected charge/mass ratio. Time selective, position accurate detection is possible using micro-channel plate (MCP) array particle multipliers followed by a phosphor screen. Two or three matched MCPs mounted in a chevron configuration provide sufficient gain to detect a single ion event. The phosphor layer is often deposited on one end of a thick fiber optic bundle that also serves as the vacuum window of the detector. The spatial resolution of such a detector assembly is $\sim 50\,\mu$m with the final resolution determined by the channel spacing and pore size of the MCP. A single ion hitting the front surface of the first MCP creates a flash of light at the phosphor screen with a diameter of 100–200 μm. The intensity of the light flash is determined by the detector gain settings and by the energy of impact of the ion. A noticeable fall-off in detector sensitivity is seen in our apparatus for heavy ions when using a repeller plate voltage of < 1000 V. The type of phosphor used determines the length of the light flash. Phosphors with short decay times are useful if TOF mass spectra are needed or for time-resolved detection. These fast phosphors, however, are usually not well matched to the colour sensitivity of the CCD camera used to record the image.

A CCD camera records light from the phosphor screen, as illustrated in Fig. 2.22. Imaging the phosphor screen onto the CCD chip is done with a commercial camera lens with a c-mount. Room light is blocked from the collection zone and a photomultiplier tube is mounted to view the entire detector signal. This PMT signal,

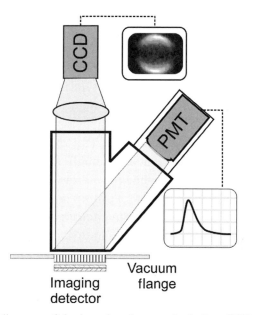

Fig. 2.22. Schematic diagram of the imaging detector including CCD camera plus readout and photomultiplier tube (PMT) connected to an oscilloscope to observe the integrated signal.

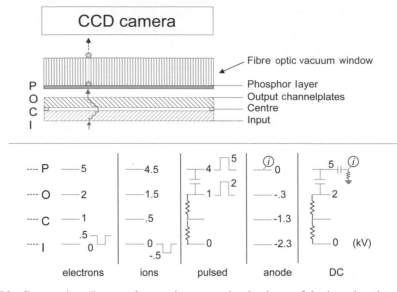

Fig. 2.23. Connection diagram for continuous and pulsed use of the imaging detector. Ions impinge on the input face of the first MCP. A centre tap connects the two MCPs. The electron cascade output of the second MCP is accelerated to the phosphor screen, which is deposited on the vacuum side of a fiber optic vacuum window. Several different schemes for connecting the channelplates are discussed in the text.

monitored on a simple oscilloscope, is very useful for the initial optimization of the imaging signal.

2.4.5.6 *Time gating of the imaging detector*

Figure 2.23 shows a schematic diagram of a typical imaging detector along with possible settings for pulsed or DC (time-of-flight) detection modes when using phosphors with a long (µs–ms) lifetime. Two approaches for time gating are shown; one uses the centre tap between the two MCPs ('electrons', 'ions') while in the other approach the voltage of the whole assembly is selected ('pulsed'). In both approaches the gain of the detector while the pulser is off is very low. In the centre tap modes the centre position and output positions are kept at a fixed voltage while the input face is pulsed to the desired voltage difference when the ion of interest arrives. A lower pulsing voltage is possible in this approach but it is necessary to ensure that in the 'on' condition the voltages on the two MCPs are equal. A slight change in the ion arrival time also results from voltages on the front plate but this effect is usually negligible. In the 'anode' mode the phosphor screen is used as a collection anode to measure the full ion TOF spectrum.

A capacitor connects the output face and phosphor screen in the 'pulsed' and 'DC' modes of Fig. 2.23. This maintains the potential difference between these two plates when the output face is pulsed to the high gain voltage. The TOF spectrum is read through a second capacitor in the 'DC' mode. In the 'pulsed' mode the front surface of the input channel plate stays at 0 V for both electrons and cations. Since both the front and back MCPs are pulsed up only at the proper time, the amount of 'leak-through' of large signals at other masses is also decreased compared to the centre-tap mode.

2.4.5.7 *Event counting*

Our ions have now been converted to light flashes on the phosphor screen. A CCD camera records this packet of light flashes for each laser shot. In the original approach, the image is integrated over many laser shots, by gathering signal intensity of the CCD chip. The summation is then done either directly on the chip (by setting a long exposure time) or afterward in the computer, or a combination. This approach (integration mode) has proven to be good, simple and direct for high ion count rates and images with inherently broad KER distributions. However, for low ion count rates and sharp structures in the velocity distribution, this approach is not the best. Here, event counting [22] provides a superior method.

Let us first discuss the topic of resolution – keeping in mind that the detector spatial resolution is linked to the speed resolution thus KER resolution. The spatial resolution of the detection system is determined by the characteristics of the MCP/phosphor/CCD system used. For a double chevron-stacked MCP detector with 10 μm-diameter channels, 12 μm centre-to-centre spacing, channel length/diameter ratio 40 : 1, and operated at 800 V/MCP, a single ion results in a $\sim$ 150 μm-diameter light flash at the phosphor screen. Each event is imaged onto the CCD chip, optically coupled by camera lens(es) or fibre optic taper(s). Mostly, each event covers many pixels of the chip – dependent on the chip dimensions and optical (de)magnification. Therefore, in the *integration mode* the resolution is limited by the ion spot size on the phosphor screen. However, since each event covers many pixels, it is possible to determine its position much more accurately by centroiding every event in each image using a fast computer algorithm. Experimentally the ultimate resolution defined by the MCP channel spacing can be retained with this centroiding technique. This is a key characteristic in *event counting*.

Now each image consists of events (which can be counted) and noise (which we want to get rid of). The computer algorithm must know what is noise and what is signal that belongs to an event. In event counting, this is achieved by defining a threshold value below which the pixel intensity is considered as noise, and above which it is considered belonging to an event. By checking surrounding pixels, the position of local maximum intensity is found, and the centroided position determined across a predefined area around this local maximum by intensity weighing. This approach thus effectively *counts* the number of events per image and stores their accurate positions. As a result, the resolution is enhanced dramatically, the detection is essentially noise free, and detector inhomogeneities are corrected for. The last point is a direct result of counting events instead of adding intensities, which generally vary enormously between different events. And do these calculations lead to data losses? No, with present-day computer technology there is sufficient time to read out the CCD array and process the image during the experiment on a real-time basis, with centroiding at sub-pixel accuracy. It must be noted that details about reading out and processing the CCD output vary greatly with the type of camera used, and although they are becoming faster, the duty cycle is mainly limited by the camera readout time and not the event counting routine.

Event counting is a clever and useful tool but it must be used wisely. First, the choice of the threshold value must be chosen such that all events are detected (otherwise it leads to count losses), and camera noise is not above this value (leading to false counts). These considerations plead also for a clear discrimination between signal and noise. The MCP voltage should be chosen high enough to obtain a clear discrimination, but not too high in order to avoid a size increase of the flash spot on the phosphor screen. If the CCD camera is not sensitive enough, one can apply an image intensified camera, but it must be noted that every amplifying device decreases the resolution. Better is to use a sensitive CCD camera, whose characteristics match well with the emission characteristics of the phosphor screen.

A next source of possible false counts is if each event covers many pixels, where the algorithm can find more than one local maximum. In this case, smoothing prior to the counting routine could help, or simply decreasing the number of pixels by a proper binning or different camera zooming. From practical experience, a 3×3 or maximum 5×5 area for each event is optimal for centroiding nicely without false-counting local sub-maxima.

When too many events are produced per laser shot, overlapping events are not recognized as multiple events, which again leads to count losses. This undercounting of events leads to false statistics (distorted speed distributions and β parameters) and thus needs to be avoided. Low count rates can be obtained experimentally by reducing molecular beam and laser intensities. Depending on ionization volume and ion mass, it is also necessary to keep the number of charged particles low enough to avoid space charge effects. Sometimes this is quite impossible, e.g., when the channel selectivity needs to be maximized with respect to competing ionization and dissociation processes in the same molecule or background molecules. Also, for competing channels in forbidden transitions, or several competing one- and multiphoton processes sometimes do not allow these signal reduction measures. In these cases where the number of ions is dictated to be high by the experiment itself, the integration mode provides the better approach. More details on the effects of event counting in the image analysis stage are presented in Chapter 3.

2.4.6 Conversion of images back to Newton spheres

We now enter the final step in the imaging experiment – the conversion of the raw output of the CCD camera back to a 2-D slice through the 3-D Newton sphere. This involves a mathematical algorithm and is discussed in detail in Chapter 3. Furthermore, the choice of experimental photodissociation/photoionization polarization and beam geometry to extract the maximum amount of information from the image has been described by Ratkitzis *et al.* [16] and in Chapter 4. In this section only a few experimental comments about this conversion will be given.

2.4.6.1 Avoiding the crush – slicing methods in imaging

Up to now we have praised the virtue of measuring the entire Newton sphere in one sweep. For many experiments this is not necessary – we only need a slice through the azimuthally symmetric 3-D sphere. If the detection conditions are such that the image measured is azimuthally symmetric, so that the normal inverse Abel transformation can be used, then crushing the 3-D sphere to catch all the data in the 2-D image is advantageous. This large amount of data provides higher signal–noise and resolution output from the transformation program. As mentioned in Section 2.2.7, however, the presence of a three-vector correlation can remove the azimuthal symmetry *for a subset of k' vectors*, and this detail makes information from the Abel transformation invalid (e.g., Fig. 2.5). For many photofragment products alignment effects represented in the three-vector correlation are either not present or are lost due to stray fields in the laboratory. Such product alignment effects should always be considered and experimentally investigated (by varying the polarization of the photoionization laser) before trusting the 2-D to 3-D transformation.

For many situations indirect (forward) convolutions can safely extract the angular information from the crushed 2-D image when a three-vector correlation is present. As discussed in Section 2.2.7, these correlations are themselves very interesting but experimentally they tend to be obscured by the 3-D to 2-D crush.

There are alternatives to crushing the 3-D Newton sphere, as discussed in more detail in Chapters 6, 8 and 11. First, we could use the Doppler effect to slice out only the in-plane (molecular beam – laser or molecular beam) species. This is illustrated in Fig. 2.24 and used by Chandler and coworkers to obtain the type of data shown in Fig. 2.5. One-photon ionization is necessary in this case in order to match the probe volume efficiently with the collision volume.

Another option is to slice the sphere in the ionization step by focusing the ionization laser into a ribbon of light with a cylindrical lens and firing the laser at time delay long enough to allow the Newton sphere to sufficiently expand. Tuning of the laser over the Doppler profile is still necessary in this method. Suzuki and coworkers

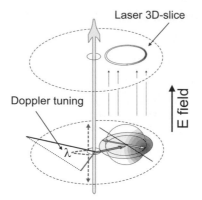

Fig. 2.24. Doppler slicing of the 3-D Newton sphere. The ionization laser propagates through the scattering plane of the crossed molecular beams and the wavelength is tuned to ionize only those molecules moving perpendicular to the laser beam direction (Doppler shift zero).

have demonstrated this approach for $(1 + 1)$ and $(2 + 1)$ REMPI detection using ion imaging [23].

A third option is to ionize the full Newton sphere and use a delayed electric field pulse or a lower constant acceleration voltage to let the neutral sphere expand before projection. This is equivalent to time-lag focusing with Wiley–McLaren TOF ion optics. Kitsopoulos and coworkers are developing this approach with considerable success [24] and it is described in more detail in Chapter 11. The centre part of the ion sphere is selected by gating the detector.

2.5 The easy experiment

Our tour of the velocity map imaging experiment in this chapter has come to an end. To help you get started in your own experiment a few hints for obtaining the easiest signal are given. When all goes wrong in our experiment we just say NO. Nitric oxide, pure or seeded in a rare gas mixture, is relatively inert and exceptionally easy to ionize at 226.167 nm in a pulsed molecular beam. We use the detector in the DC mode (Fig. 2.23) and look for ions. The first step is to optimize the timing of the molecular beam–laser beam overlap. A few micro-Joules of unfocused laser light tuned to transitions from low rotational states of NO (for a cold molecular beam) is sufficient to create an ion signal even when the molecular beam time delay is incorrect. The dot becomes much stronger when the proper time delay is found. $(1 + 1)$ REMPI photoionization takes place and the NO^+ signal will appear as a dot at the centre of the detector. The electrostatic lens is tuned to form the smallest dot. With a fixed repeller voltage the optimal extractor voltage (within 1 volt!) produces

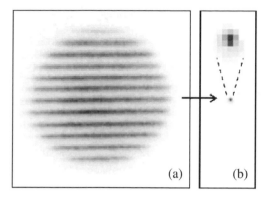

Fig. 2.25. Velocity map image of cold NO molecules photoionized with 226 nm light (1 + 1) REMPI, with detection of NO^+. The electrostatic lens extractor/repeller ratio is set lower than that for velocity mapping. This causes magnification of zero-velocity species. In panel (a) a series of images are added together, each with the laser beam displaced in small steps across the molecular beam diameter. This helps in finding the centre of the molecular beam and also the proper distance for focusing the REMPI laser. In panel (b) the electrostatic lens is set for velocity mapping and the image collapses to almost a single camera pixel.

the smallest possible dot. The detector can then be set in the pulsed mode and the timing optimized for the NO^+ dot. The time delay for pulsing the detector is sensitive to the position of the laser beam between the repeller and extractor plate. Once this position is chosen, we centre two irises on the beam before and after the chamber. These become 'untouchable' definers of the laser beam direction.

We then manipulate our steering optics (Fig. 2.21) to centre the signal dot on the centre of the detector. After finding the extractor voltage for velocity mapping we then set the extractor ~ 200 V *lower*, so that the electrostatic lens operates as a microscope for ions of zero KER. Figure 2.25 shows a summation image where the laser beam is displaced stepwise through the NO molecular beam.

The largest diameter stripe cuts through the centre of the molecular beam. When using a focusing lens in the laser beam at this position, the stripe will be brightest in the centre when the focal point of the laser is centred on the molecular beam. We then return to the velocity mapping voltage and reverse the polarity of the electrostatic lens for photoelectron detection (Fig. 2.23). At short time delays an electron image appears with a $\cos^2\theta$ distribution when the laser polarization is set vertical (parallel to the detector face). If necessary the electron image can be re-centred on the detector using the steering device.

You're now ready to velocity map cations, anions, or electrons (don't forget to switch the bias on the ion lens when switching from positively to negatively charged particles). Good luck!

References

1. D. W. Chandler, P. L. Houston, *J. Chem. Phys.* **87**, 1445 (1987).
2. A. T. J. B. Eppink, D. H. Parker, *Rev. Sci. Instrum.* **68**, 3477 (1997).
3. R. D. Levine, R. B. Bernstein, *Molecular Reaction Dynamics and Chemical Reactivity*. (Oxford University Press, New York, 1987).
4. R. F. Delmdahl, B. L. G. Bakker, D. H. Parker, *J. Chem. Phys.* **113**, 7728 (2000).
5. Wiley Series in Ion Chemistry and Physics: *High Resolution Laser Photoionization and Photoelectron Studies*, eds. I. Powis, T. Baer, C. Y. Ng, (John Wiley & Sons, New York, 1995).
6. D. H. Parker, R. B. Bernstein, *Annu. Revs. Phys. Chem.* **40**, 561 (1989).
7. D. A. Case, G. M. McClelland, D. R. Herschbach, *Mol. Phys.* **35**, 541 (1978).
8. K. T. Lorenz, D. W. Chandler, J. W. Barr, W. Chen, G. L. Barnes, J. I. Cline, *Science* **293**, 2063 (2001).
9. A. T. J. B. Eppink, D. H. Parker, M. H. M. Janssen, B. Buijsse, W. J. van der Zande, *J. Chem. Phys.* **108**, 1305 (1998).
10. D. M. Sonnefroh, K. Liu, *Chem. Phys. Lett.* **176**, 183 (1991).
11. M. Drabbels, C.G. Morgan, D. S. McGuire, A. M. Wodte, *J. Chem. Phys.* **102**, 611 (1995); C. G. Morgan, M. Drabbels, A. M. Wodte, *J. Chem. Phys.* **105**, 4550 (1996).
12. G. Herzberg, *Molecular Spectra and Molecular Structure I: Spectra of Diatomic Molecules*, (Krieger Publishing Company, Florida, 1989).
13. C. Jonah, *J. Chem. Phys.* **55**, 1915 (1971).
14. A. J. Alexander, R. N. Zare, *Accts. Chem. Res*, **33**, 199 (2000).
15. D. H. Parker, R. Delmdahl, B. Bakker, H. P. Loock, *J. Chin. Chem. Soc.,* **48**, 1 (2001).
16. T. P. Rakitzis, P. C. Samartzis, T. Kitsopoulos, *J. Chem. Phys.* **111**, 10415 (1999).
17. E. Wrede, S. Laubach, S. Schulenberg, A. Brown, A. J. Orr-Ewing, M. N. R. Ashfold, *J. Chem. Phys.* **114**, 2629 (2001).
18. H.-P. Loock, B. L. G. Bakker, D. H. Parker, *Can. J. Phys.* **79**, 211 (2001).
19. W. C. Wiley, I. H. McClaren, *Rev. Sci. Instrum.* **26**, 1150 (1955).
20. H. L. Offerhaus, C. Nicole, F. Lepine, C. Bordas, F. Rosca-Pruna, M. J. J. Vrakking, *Rev. Sci. Instrum.* **85**, 4024 (2000).
21. N. Yonekura, C. Gebauer, H. Kohguchi, T. Suzuki, *Rev. Sci. Instrum.* **70**, 3265 (1999).
22. B. Chang, R. C. Hoetzlein, J. A. Mueller, J. D. Geiser, P. L. Houston, *Rev. Sci. Instrum.* **69**, 1665 (1998).
23. K. Tonokura, T. Suzuki, *Chem. Phys. Lett.* **224**, 1 (1994).
24. T. N. Kitosopoulos, C. R. Gebhardt, T. P. Rakitzis, *J. Chem. Phys.* **115**, 9727 (2001).

3

Reconstruction methods

ANDRÉ T. J. B. EPPINK, SHIOU-MIN WU
AND BENJAMIN J. WHITAKER

3.1 Introduction

Charged particle imaging provides us with very beautiful pictures that offer graphic insight into chemical dynamics. Although it is often the case that general dynamical information can be deduced by simple inspection of the primary data, the images obtained in the typical imaging experiment are, in fact, projections of a three-dimensional (3-D) object onto a two-dimensional (2-D) screen. In order to extract all the information potentially available to us we need to consider what data recovery techniques are available to reconstruct the 3-D velocity distribution of the charged particles created in the experiment from the image we actually record.

There are two fundamentally different approaches; *inversion methods* and *forward convolution methods*. Inversion methods make use of the fact that if the original (3-D) distribution has an axis of cylindrical symmetry its (2-D) projection parallel to this axis contains enough information to unambiguously reconstruct the full (3-D) distribution. As we have seen in the previous two chapters, such an axis of symmetry in laboratory space can be found in many photodissociation or bimolecular scattering experiments. However, if there is no cylindrical symmetry in the experiment, a forward convolution method is generally necessary. Here, the experiment is simulated in a computer model that produces (2-D) data that are then compared with the experimental data. By iteratively optimizing parameters in the computer model the best reconstruction of the experimental data is sought.

This chapter mainly focuses on inversion methods. Various ways in which these may be implemented are described, and a comparison between five different algorithms is presented using both simulated and experimental input images. At the end, the forward convolution method is also described.

3.2 Symmetric distributions

As we have seen previously, in many cases the velocity distribution of the charged particles has a well-defined axis of symmetry. For example, in a photodissociation experiment we can arrange the polarization of the photolysis laser to lie parallel to the imaging plane so that the photofragments are ejected with cylindrical symmetry about an axis parallel to the detector face. Under these circumstances a mathematical procedure such as the well-known inverse Abel transform [1] is able to recover this 3-D velocity from its projection onto a 2-D detector.

In a typical imaging experiment, a probe laser is used to state-selectively ionize a sub-set of the photofragments at some given time after the primary photolysis event. The photofragments in the selected quantum state now exist as ions on the surfaces of a set of nested spheres. The diameter of each sphere at any given moment after the photolysis laser pulse is proportional to the square root of the kinetic energy released into that particular channel. There may be a number of these because even though the quantum state of the particular fragment to be imaged has been fully determined, the quantum state of its invisible partner has not. However, because the internal energy of the unimaged partner is quantized, conservation of energy dictates that the speed distribution of the probed fragment is also quantized. The nested spheres of ions are then projected by means of an electric field onto the imaging detector and recorded. Ignoring possible complications that can arise if the photofragments are aligned with respect to the symmetry axis, and which are dealt with later in the following chapter by Rakitzis, the projected image contains enough information for us to be able to recover the 3-D angular distribution for each of the shells of ions that are summed together in the image.

Figure 3.1 illustrates this for the case of photolysis of Br_2 at three different wavelengths. Two channels are visible. The larger (faster) ring is due to Br $(^2P_{1/2})$ atoms produced in coincidence with other ground state Br atoms. The inner (slower) ring is due to Br $(^2P_{1/2})$ produced in coincidence with excited state Br $(^2P_{3/2})$ atoms. It is also clear from the image that the two channels have different angular distributions and that the branching ratio between these channels varies with the photolysis wavelength. The angular distributions of both channels are symmetric about the direction of the photolysis laser's polarization vector but the fast Br $(^2P_{1/2})$ atoms are ejected predominantly perpendicular to the polarization vector whilst the slow atoms are mostly ejected along the polarization direction. The dynamical explanation for this behaviour has been described in detail by Wrede *et al.* [2].

A more complicated example is shown in Fig. 3.2 which shows primary data obtained for the O^+ photoions obtained by $(2 + 1)$ resonantly enhanced multiphoton ionization (REMPI) of O_2. A large number of possible channels exist for this process

Photolysis wavelength (nm)

400 460 480

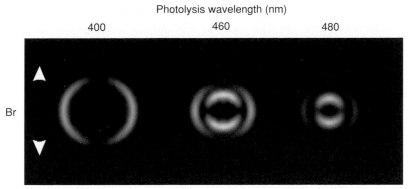

Br

Fig. 3.1. Image of $Br^+(^3P_0)$ produced by resonant ionization of ground state Br $(^2P_{1/2})$ atoms following the photolysis of Br_2 at 400, 460 and 480 nm. The larger (faster) ring is due to Br $(^2P_{1/2})$ atoms produced in coincidence with other ground state Br atoms. The inner (slower) ring is due to Br $(^2P_{1/2})$ produced in coincidence with excited state Br $(^2P_{3/2})$ atoms. The data were obtained by E Wrede *et al.* and are discussed in more detail in Ref. [2].

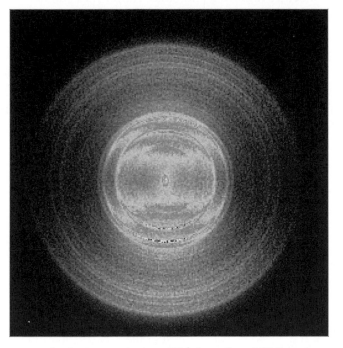

Fig. 3.2. Event counted velocity map image of O^+ from O_2 at 225.0 nm (see text for details).

since the 225 nm light used in these experiments is (2 photon) resonant with the $3d\pi(^3\Sigma_{1g})$ Rydberg state of molecular oxygen which can then absorb the third laser photon to create a superexcited O_2^* molecule which can suffer a variety of fates; directly or auto-ionizing to yield $O_2^+(X^2\Pi_g)$ or $O_2^+(a^4\Pi_u)$ in a number of different vibrational levels, or dissociating into neutral oxygen atoms in a variety of electronic states. Products in both molecular ion channels and atomic channels can then absorb a fourth photon to yield both $O^+(^4S)$ and $O^+(^2D)$ as the finally detected products in the ion image. It is a remarkable testimony to the resolution of velocity map imaging that this plethora of outcomes can be disentangled and every feature in the image shown in Fig. 3.2 assigned and characterized [3].

3.3 Abel and Hankel inversion methods

Although we can make qualitative deductions about the controlling dynamics from the raw images, clearly we would do better if we could extract the true velocity distribution of the photofragments from their projected image. Because we can assume that the velocity distribution of the ion cloud before projection was cylindrically symmetric we are, in fact, able to back-transform the image to recover the original distribution essentially exactly. Figure 3.3a illustrates the form of the ion cloud for a single shell of velocities following photodissocation via a parallel transition. In this case, the majority of the photofragments are ejected along the line defined by the photolysis polarization according to a $\cos^2\vartheta$ distribution, where ϑ is the angle between $\mathbf{E}$, the polarization vector, and $\mathbf{v}$, the photofragment velocity vector. We can write the velocity distribution of the photofragment ions in Cartesian coordinates centred on the point of photolysis as $i(x, y, z)$. Take z to be the symmetry axis (i.e. the laser polarization vector), and imagine that an extraction field in the y direction accelerates the ions. Assuming that the ion optics have been adjusted so that the entire ion cloud arrives within the gate-width of the imaging system, the measured distribution on the detector, shown in Fig. 3.3b, is:

$$p(x, z) = \int_{-\infty}^{\infty} i(x, y, z)dy. \tag{3.1}$$

If we consider just one row of the image, say $f(x; z_0)$, taken along the x axis at some value of $z = z_0$, we have:

$$f(x) = p(x, z_0) = \int_{-\infty}^{\infty} s(x, y)dy = 2\int_{0}^{\infty} s(x, y)dy, \tag{3.2}$$

where $s(x, y) = i(x, y; z_0)$ is a slice through the 3-D distribution perpendicular to the symmetry axis taken at z_0. The function $f(x)$ is illustrated for a slice taken at $z = 0.9$ through the projected image in Fig. 3.3c. Notice that the function does not rise vertically at the edge of the image. This is because in order to construct Fig. 3.3b. it was assumed that the speed distribution of the state-selected photofragments was described by a narrow Gaussian function rather than a delta

(a) (b)

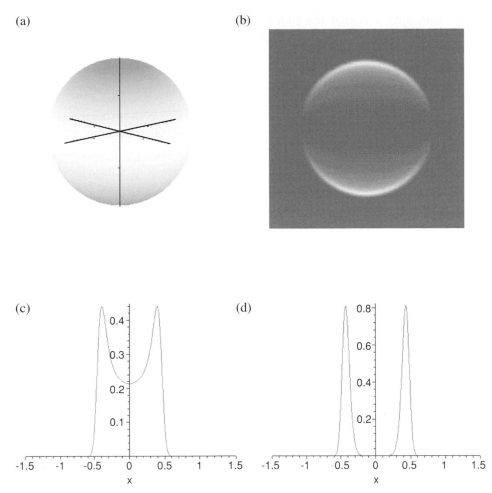

(c) (d)

Fig. 3.3. (a) A representation of cylindrically symmetric distribution of state-selected photofragments produced in a photofragmentation experiment. In a typical experiment the fragments will be produced in coincidence with a number of 'invisible' partners so a number of concentric fragment shells will be created, of which only one is illustrated here. (b) The projection of a cylindrically symmetry ion cloud depicted in (a) onto a two-dimensional screen. (c) A typical slice through the projected image. (d) The true speed distribution slice through the original velocity distribution corresponding to (c).

function, and reflects the fact that the energy resolution of an imaging experiment is never perfect. In the figure it is assumed that the projected speeds can be resolved to one part in a thousand, which is marginally better than can be achieved in practice. In order to reconstruct the original velocity distribution (illustrated in Fig. 3.3d), our job is somehow to recover $s(x, y)$ from the measured function $f(x)$.

Our assumption is that the function, $s(x, y)$ is cylindrically symmetric and so we can express (3.2) in polar coordinates as:

$$f(x) = 2 \int_x^\infty \frac{s(r)r}{\sqrt{r^2 - x^2}} dr \qquad (3.3)$$

by a simple change of variables with $r^2 = x^2 + y^2$. This is the Abel transform and it arises frequently in image-processing applications when circular symmetry is present [4]. The inverse transform, which will recover $s(r)$ from the measurement of $f(x)$, can be found by applying the Fourier transform convolution theorem. It is given by [5]:

$$s(r) = \frac{1}{\pi} \int_r^\infty \frac{df/dx}{\sqrt{x^2 - r^2}} dx. \qquad (3.4)$$

This can then be used to recover the original distribution, since we now have a prescription for obtaining $s(r; z_0)$ for each row, $f(x; z_0)$, of the projected image.

Solving equation (3.4) is difficult in practice because of the singularity at $r^2 = x^2$, and because the derivative in the integrand tends to magnify noise. The application of the Abel transform, formulated above, also assumes that the input data are symmetric. Because of noise and effects such as mismatch between the ionizing laser bandwidth and the Doppler profile of the fragments, it is frequently the case that real images exhibit some degree of asymmetry. One method of performing the inversion, proposed by Smith and Keefer [6], removes some of these difficulties. It involves taking the Fourier transform of equation (3.2):

$$F\{f(x)\} = \int_{-\infty}^\infty \int_{-\infty}^\infty s(\sqrt{x^2 + y^2}) \exp(-2\pi ixq) dx dy \qquad (3.5)$$

and then noticing that the definition of the zero-order Bessel function of the first kind, J_0, is given by:

$$J_0(z) = \frac{1}{2\pi} \int_0^{2\pi} \exp(-iz \cos \vartheta) d\vartheta. \qquad (3.6)$$

Changing to polar coordinates and substituting this into equation (3.5) yields:

$$F\{f(x)\} = 2\pi \int_0^\infty rs(r) J_0(2\pi rq) dr. \qquad (3.7)$$

As noted by Smith and Keefer and by Castleman [7], the right-hand side of this equation is the zero-order Hankel transform of $s(r)$. Since the Hankel transform is identical to its inverse, the original distribution $s(r)$ can then be recovered by taking the Hankel transform of the Fourier transform of the projected intensity:

$$s(r) = H[F\{f(x)\}]$$

$$= 2\pi \int_0^\infty q J_0(2\pi r q) \left[\int_{-\infty}^\infty f(x) \exp(-2\pi x q) dx \right] dq. \qquad (3.8)$$

Equation (3.8) has several advantages over the Abel inversion given by (3.4) in that it avoids the difficulty associated with the lower limit of integration and allows one to filter and smooth the data in the Fourier transform step. Computationally, the method used to evaluate equation (3.8) is to apply the Cooley–Tukey fast Fourier transform algorithm [8] on each line of the image and then either to compute the Bessel function convolution by direct summation, or more efficiently, to sum selected Fourier components of the integrand according to an algorithm first given by Candel [9]. On a modern personal computer, this is straightforward. It is worth noting that if $f(x)$ is an error-free projection of a real and cylindrically symmetric object centred at $x = 0$, as originally assumed, it follows that $F\{f(x)\}$, its Fourier transform, is also a real and even function. This property can be used to check that each line of the image is correctly centred on the symmetry axis, or to correct for any slight skewing in the input image. Although the Hankel transform formulation of the image reconstruction algorithm reduces the problems associated with noise in the image, it does not completely eliminate them. In particular, because the Bessel function oscillates rapidly close to the origin, noise is magnified along the centreline of the reconstructed velocity distribution.

3.4 Back-projection and onion peeling methods

An alternative strategy is to use a back-projection approach. Here one considers explicitly the Jacobian of the co-ordinate transformation from the velocity distribution space, (r, ϑ), to that of the image plane, (x, z). In general, this is a very complicated function. However, for the case where the ratio of the electrostatic energy acquired by the ions in the acceleration region is large compared to their initial kinetic energy one can obtain a relatively straightforward analytical expression which can be used to project points on the image back to their origins in the velocity distribution one wants to recover [10,11].

We start by considering a charged particle created at a point $x = y = z = 0$ (see Fig. 3.4) and travelling with initial velocity components

$$v_x = v_0 \sin \vartheta \cos \varphi$$

$$v_y = v_0 \sin \vartheta \cos \varphi$$

$$v_z = v_0 \cos \vartheta.$$

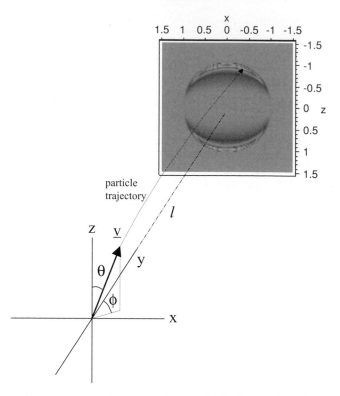

Fig. 3.4. Schematic diagram of the projection method. Charged particles are created at a point (0,0,0) in Cartesian co-ordinates with an initial velocity **v** determined by energy and momentum constraints and the molecular dynamics. Electrostatic optics project the particles along the y axis to the imaging detector.

Remembering that the particle is created in the presence of a homogeneous electric field, **F**, which accelerates it towards the detector centred at $x = z = 0$, $y = l$, and applying Newton's laws of motion, it is straightforward to calculate the co-ordinates (x, z) of the particle at the detector:

$$x = \frac{2l \cos \varphi \sin \vartheta}{\rho} \left(\sqrt{\sin^2 \varphi \sin^2 \vartheta + \rho} - \sin \varphi \sin \vartheta \right)$$

$$z = \frac{2l \cos \vartheta}{\rho} \left(\sqrt{\sin^2 \varphi \sin^2 \vartheta + \rho} - \sin \varphi \sin \vartheta \right)$$

(3.9)

where

$$\rho = \frac{qFl}{w_0}$$

(3.10)

is the ratio between the electrostatic energy acquired by the particle and its initial kinetic energy, w_0 (here q represents the charge on the particle). In obtaining

(3.9) we have assumed that the (uniform) accelerating field is present for the entire flight of the particle from its point of origin to the detector. This situation is quite often realized in photoelectron imaging experiments but rarely, if ever, in photoion imaging. However, this detail will not affect our subsequent argument since the existence of a field-free region in the flight tube merely adds a uniform magnification factor, $t\sqrt{2w_0/m}$, where t is the flight time in the field-free region.

In many imaging applications the initial kinetic energy of the particles is of order 1 eV or less, whilst the accelerating field may impart several thousand eV of kinetic energy. Consequently, the parameter ρ is likely to be large. As ρ tends to infinity equations (3.9) reduce to:

$$x = \frac{2l\cos\varphi\sin\vartheta}{\sqrt{\rho}}$$

$$z = \frac{2l\cos\vartheta}{\sqrt{\rho}}. \tag{3.11}$$

We can then invert equations (3.11) to obtain expressions for the angles describing the initial velocity components of a particle striking the detector at the point (x, z) for a given value of ρ:

$$\vartheta = \arccos\left(\frac{z\sqrt{\rho}}{2l}\right)$$

$$\varphi = \arccos\left(\frac{x\sqrt{\rho}}{2l\sin\vartheta}\right) = \arccos\left(\frac{x\sqrt{\rho}}{l\sqrt{4 - \rho z^2 l^2}}\right) \tag{3.12}$$

from which we see that, in the limit of large ρ, the z coordinate of any point in the image uniquely determines the colatitude, ϑ, of the imaged particle's initial velocity. The azimuth, φ, is a function of both coordinates, x, and z, but for a given pair of ρ and z values it only depends on x. Therefore, if there was only a single (large) value of ρ contributing to the image, i.e. if the initial speed distribution of the particles was a δ function, equations (3.12) give us means to transform the image back to the initial velocity distribution. Unfortunately, even in the simplest case of a single channel contributing to the image there will be some spread in the initial speed distribution because of the laser linewidth and the thermal distribution of states in the sample, etc.

The way forward is to consider the contribution of the most energetic particle to the final image. We then remove the contribution of this value of ρ from the image, and proceed iteratively with the remaining image until all particle speeds have been accounted for. This process is picturesquely called *onion peeling*. To develop the algorithm one needs to calculate the Jacobian of the transformation from the coordinates (ϑ, φ) to (x, z) for a given value of ρ. The initial angular distribution

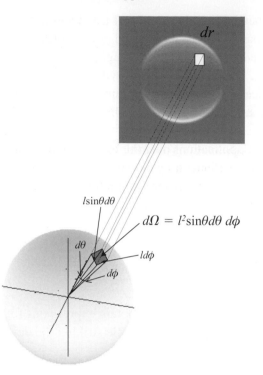

Fig. 3.5. Illustration of the transformation of the surface element $d\Omega$ to the surface element dr.

$s(\rho, \vartheta, \varphi)$ is the number of particles ejected into the solid angle Ω defined by the Euler angles (ϑ, φ). The corresponding number of particles hitting the detector at the position (x, z) is $f(\rho, x, z)$. The two distributions are related by:

$$f(\rho, x, z)dr = s(\rho, \vartheta, \varphi)d\Omega \tag{3.13}$$

where $d\Omega$ is the elemental surface on the sphere in the initial coordinates and dr is the corresponding surface element on the detector. As shown in Fig. 3.5, the surface element is $d\Omega = l^2 \sin \vartheta \, d\vartheta \, d\varphi$. The corresponding surface element on the image can be expressed in the original coordinates by $dr = |J| \, d\vartheta \, d\varphi$ where J is the Jacobian of the coordinate transformation defined by:

$$J = \frac{dx}{d\vartheta} \frac{dz}{d\varphi} - \frac{dx}{d\varphi} \frac{dz}{d\vartheta}. \tag{3.14}$$

Using equations (3.11), which we stress once again are valid only for large values of ρ, one finds that the Jacobian can be expressed as:

$$|J| = \frac{4l^2}{\rho} \sin^2 \vartheta \, |\sin \varphi|. \tag{3.15}$$

Then using equations (3.12) and (3.13) we obtain:

$$f(\rho, x, z) = s(\rho, \vartheta, \varphi) \frac{\rho}{4 \sin \vartheta \sqrt{1 - x^2/x_0^2}}, \qquad (3.16)$$

where $x_0 = \sqrt{4l^2 - \rho z^2/\rho}$. From (3.12) and the definition of x_0, it follows that $\sin \vartheta = x_0 \sqrt{\rho}/2l$ and so (3.16) gives an explicit relation between the density of particles with a given initial kinetic energy (expressed in terms of the ratio ρ) hitting the detector at the position (x, z) and the initial angular velocity distribution.

Equation (3.16) describes the forward transform from velocity distribution to image and its inverse gives us a prescription for the back-projection of the image to the velocity distribution. One notices immediately that (3.16) diverges when $x = x_0$. This is similar to the problem that occurred with the Abel transform, equation (3.4), but leads to different consequences. The parameter x_0 represents the maximum value of x that a particle with initial kinetic energy w_0 (or, alternatively expressed, acceleration ratio ρ) can reach for a given line, z, across the image. This follows by considering the inversion of equations (3.9) which yield the general expressions [10, 11]:

$$\varphi^{\pm} = \arctan \left(\frac{-2 \pm \sqrt{4(\rho + 1) - \left[\rho^2 \left(x^2 + z^2 \right) / l^2 \right]}}{\rho x l} \right)$$

$$\qquad (3.17)$$

$$\vartheta^{\pm} = \arctan \left(\frac{x}{z \cos \varphi^{\pm}} \right)$$

for arbitrary values of ρ. For φ to be real the argument under the square root in (3.17) must be positive which implies that:

$$\sqrt{x^2 + z^2} \leq 2l \frac{\sqrt{\rho + 1}}{\rho} \qquad (3.18)$$

in general, or that $x \leq \sqrt{4l^2 - \rho z^2/\rho}$ for the limiting case of large ρ that we are considering here. Clearly, (3.16) is only defined for $|x| \leq x_0$.

For continuous distributions, the back-projection from the measured distribution, $f(\rho, x, z)$, to the initial distribution of velocities, $s(\rho, \vartheta, \varphi)$, using (3.16) is exact (within the assumption of large ρ), however, pixelation of the data introduces approximations into the inversion. This has important consequences which we explore further below, but the basic idea behind the inversion procedure is to start with the most energetic particle in the image. This will be found at the outer edge of the recorded image along the line $x = 0, z$. Using (3.18) we know the value of the energy ratio, ρ, for this particle. This is equivalent to knowing the particle's initial kinetic energy. Then, from (3.16), we can work out the contribution that particles with this energy will make to the rest of the image and subtract this from the image.

is repeated until the differences between the calculated and the experimental
2-D projections are acceptably small. Since this method does not treat the inversion line-by-line but rather uses the radial information, the procedure obviates
the centre-line noise problems which appear in other methods. Instead, the noise is
projected towards the centre of the image.

The inversion algorithm is developed as follows. First, one notes that the 3-D
velocity distribution can be written as the product of a radial distribution $P_1(r)$ and
a radial-dependent angular distribution $P_2(r, \vartheta)$:

$$P(x, y, z) = P(r, \vartheta, \varphi) = P_1(r)P_2(r, \vartheta) \tag{3.26}$$

where we simply put the velocity coordinates equal to the pixel positions, i.e. $x = v_j$
etc. Implicitly the distribution is assumed to be cylindrically symmetric, thus there
is no dependency on the azimuthal angle φ. To be consistent with the notation
of the original publication and in contrast to the convention we have adopted up
to now, we will take the image to lie in the x–y plane with the symmetry axis
lying parallel to the y direction. So, against the normal convention, the polar angle
ϑ is defined by the angle between a vector $\mathbf{r}$ and the y axis, and the z axis is
the projection coordinate, lying along the detector (time-of-flight) axis. The 3-D
velocity distribution is chosen to be normalized, thus:

$$2\pi \iint P_1(r)r^2 P_2(r, \vartheta) \sin \vartheta \, d\vartheta \, dr = 1. \tag{3.27}$$

The 2-D experimental distribution can be written as a product of two
functions:

$$Q_{\exp}(x, y) = Q_{\exp}(r, \alpha) = Q_{1,\exp}(r)Q_{2,\exp}(r, \alpha), \tag{3.28}$$

with a similar normalization condition:

$$\iint Q_{1,\exp}(r)r Q_{2,\exp}(r, \alpha) d\alpha dr = 1. \tag{3.29}$$

In the iterative procedure, the experimentally observed radial and angular distributions of the image are used as the first guess for the velocity radial and angular
functions, thus:

$$\begin{aligned} P_{1,i=0}(r) &= Q_{1,\exp}(r)/2\pi r, \\ P_{2,i=0}(r, \vartheta) &= Q_{2,\exp}(r, \alpha = \vartheta) \end{aligned} \tag{3.30}$$

where i is used as the iteration index. With this Ansatz, a new 2-D image is calculated from which new radial and angular distributions $Q_{1,i=0}(r)$ and $Q_{2,i=0}(r, \alpha)$
are obtained. These are compared to the experimental functions, and a correction

applied according to:

$$P_{1,i}(r) = P_{1,i-1}(r) - c_1[Q_{1,i-1}(r) - Q_{1,\exp}(r)]/2\pi r,$$
$$P_{2,i}(r, \vartheta) = P_{2,i-1}(r, \vartheta) - c_2[Q_{2,i-1}(r, \alpha = \vartheta) - Q_{2,\exp}(r, \alpha = \vartheta)]. \qquad (3.31)$$

This procedure is repeated iteratively until reasonable agreement between the calculated and experimental data is achieved. The parameters c_1 and c_2 determine the magnitude of the corrections that are applied to the functions, typical values are 2.0 and 1.0 respectively, and an acceptable level of convergence is generally achieved after 10–20 iterations.

As an alternative basic approach, the experimental image can be inverted with the Fourier–Hankel method prior to the iterative procedure. This option can speed up the convergence for images that exhibit good signal-to-noise. More details on the forward projections used in the algorithm to produce the 2-D data from the 3-D velocity distributions can be found in Vrakking's original paper [13].

3.6 Basis set expansion or BASEX method

A similar approach to Vrakking's method has been proposed by Dribinski, Reisler, and co-workers [14] in which the experimental projection data are expanded over a set of basis functions that are analytical projections of known well-behaved functions (very similar to Gaussian functions). The expansion coefficients then yield directly the information necessary to reconstruct the 3-D velocity distribution. The method is called BASEX (from BAsis Set EXpansion).

The cylindrically symmetric 3-D velocity distribution which we want to reconstruct can be described with a function which depends on two coordinates, $I = I(r, z)$, where reverting to our traditional convention the z axis corresponds to the symmetry axis. The projection of this function onto the (x, z) detector plane is described by the Abel integral (see (3.3)):

$$P(x, z) = 2 \int_{|x|}^{\infty} \frac{rI(r, z)}{\sqrt{r^2 - x^2}} dr. \qquad (3.32)$$

In the imaging experiment, this projection function is binned on the CCD chip, which has a size of $N_x \times N_z$ pixels, leading to the digitized projection function $\mathbf{P}$

$$\mathbf{P}_{ij} = 2 \int h(x - x_i, z - z_i) dx dz \int_{|x|}^{\infty} \frac{rI(r, z)}{\sqrt{r^2 - x^2}} dr, \qquad (3.33)$$

where $h(x, z)$ defines an instrumental function. The idea is now to make an expansion of this function in terms of a basis of known projection functions.

Assume we have a basis set of functions $\{f_k(r, z)\}$ for $(k = 0, \ldots, K - 1)$ and their corresponding projection basis set $\{\mathbf{G}_k\}$. They are related through:

$$\mathbf{G}_{kij} = 2 \int h(x - x_i, z - z_i) dx dz \int_{|x|}^{\infty} \frac{r f_k(r, z)}{\sqrt{r^2 - x^2}} dr. \tag{3.34}$$

If the basis set is appropriate for making a complete representation of $\mathbf{P}$, and if the basis set functions are well-behaved under the Abel transform – preferably such that the Abel integral is mathematically exactly solvable – we may describe the 3-D velocity distribution and its projection as expansions in the basis set, using the same expansion coefficients:

$$I(r, z) = \sum_{k=0}^{K-1} C_k f_k(r, z),$$

$$\mathbf{P}_{ij} = \sum_{k=0}^{K-1} C_k \mathbf{G}_{kij} \tag{3.35}$$

or, in matrix form $\mathbf{P} = \mathbf{CG}$ with the coefficients vector $\mathbf{C} = (C_0, \ldots, C_{K-1})$ and the projection basis matrix $\mathbf{G} = (\mathbf{G}_0, \ldots, \mathbf{G}_{K-1})^{\mathrm{T}}$. Now the coefficients which are sought for, can be found from the Tikhonov regularization method [15]:

$$\mathbf{C} = \mathbf{PG}^{\mathrm{T}}(\mathbf{GG}^{\mathrm{T}} + q^2 \mathbf{I})^{-1}, \tag{3.36}$$

where $\mathbf{I}$ is the identity matrix, and q is a regularization parameter. The number of basis functions, K, is ideally chosen equal to the number of data points, $N_x \times N_z$. However, for typical CCD arrays, this would result in a value of K in the order of 10^5–10^6, which makes the inversion problem numerically unmanageable. However, as we have already seen, the problem is separable into two independent variables, so that we can write:

$$I(r, z) = \sum_{k=0}^{K_x-1} \sum_{m=0}^{K_z-1} C_{km} \rho_k(r) \zeta_m(z),$$

$$\mathbf{P}_{ij} = \sum_{k=0}^{K_x-1} \sum_{m=0}^{K_z-1} C_{km} \mathbf{X}_{ki} \mathbf{Z}_{mj}, \tag{3.37}$$

or, in matrix form $\mathbf{P} = \mathbf{X}^{\mathrm{T}} \mathbf{CZ}$, with:

$$\mathbf{X}_{ki} = 2 \int h_x(x - x_i) dx \int_{|x|}^{\infty} \frac{r \rho_k(r)}{\sqrt{r^2 - x^2}} dr,$$

$$\mathbf{Z}_{mj} = \int h_z(z - z_j) \zeta_m(z) dz. \tag{3.38}$$

This reduces the basis-set to a product basis-set of size $K_x \times K_z$. Now the expansion coefficients are found from $\mathbf{C} = \mathbf{APB}$, with $\mathbf{A} = (\mathbf{XX}^{\mathrm{T}} + q_1^2 \mathbf{I})^{-1} \mathbf{X}$ and

$\mathbf{B} = \mathbf{Z}^T(\mathbf{Z}\mathbf{Z}^T + q_2^2\mathbf{I})^{-1}$. Because the matrices $\mathbf{A}$ and $\mathbf{B}$ are independent of the experimental projection matrix $\mathbf{P}$, they can be computed once and used afterward for the retrieval of the expansion coefficients $\mathbf{C}$ by relatively fast matrix multiplication.

As already noted, the basis-set functions $\{\rho_k(r)\}$ and $\{\zeta_m(z)\}$ are optimal if their projection functions can be expressed analytically and if they are complete in the sense that any projection matrix can be expanded in this basis. They should therefore be able to account for sharp features of the order of one pixel and be smooth on a smaller scale. In BASEX, the following radial basis functions have been chosen that satisfy these conditions:

$$\rho_k(r) = (e/k^2)^{k^2}(r/\sigma)^{2k^2} \exp[-(r/\sigma)^2] \tag{3.39}$$

for $k = (0, \ldots, K_x - 1)$, $K_x \leq (N_x + 1)/2$, where σ is of the order of the distance between the pixels. These functions have their maximum at $r = k\sigma$ and are practically indistinguishable from Gaussian functions, i.e. $\rho_k(r) \approx \exp[-2(r/\sigma - k)^2]$, for sufficiently large k.

Now the matrix $\mathbf{X}$ is solvable exactly, and yields, assuming an instrumental function $h_x(x) = \delta(x)$:

$$\mathbf{X}_{ki} = 2\sigma\rho_k(x_i)\left[1 + \sum_{l=1}^{k^2}(x_i/\sigma)^{-2l}\prod_{m=1}^{l}\frac{(k^2 + 1 - m)(m - \frac{1}{2})}{m}\right]. \tag{3.40}$$

The basis functions along the z-axis are chosen to be the same as the radial basis functions, thus $\zeta_m(z) = \rho_m(z)$, which leads to $\mathbf{Z}_{mj} = \rho_m(z_j)$ for $m = (0, \ldots, K_z - 1)$, $K_z \leq (N_z - 1)/2$. In the present implemenation of BASEX, for an image of 1001×1001 pixels, two sets of basis functions have been used: (i) 'broad' functions with $\sigma = 2$, $K_x = K_z = 226$, $q_1^2 = 50$, $q_2^2 = 0$, and (ii) 'sharp' functions with $\sigma = 1$, $K_x = K_z = 251$, $q_1^2 = 50$, and $q_2^2 = 0$.

After the inversion, speed and angular distributions can be extracted from the reconstructed image. 'Traditionally', the speed distribution is obtained from the reconstructed velocity distribution:

$$P(v) = \frac{1}{(P(v))_{max}}\int_0^{\pi}I(v, \vartheta)v\sin\vartheta\,d\vartheta, \tag{3.41}$$

where $v = r$ and ϑ is found from $z = r\cos\vartheta$. In pixelated form this becomes:

$$P(v) = \frac{1}{(P(v))_{max}}\sum_{n=0}^{v}I\left(v, \frac{\pi n}{2v}\right)v\sin\left(\frac{\pi n}{2v}\right), \tag{3.42}$$

where $v = (1, \ldots, v_{max})$ and $I(v, \pi n/2v)$ is calculated from the reconstructed image as an interpolated value between the four surrounding pixels. Alternatively, the BASEX method offers the possibility of calculating the speed distribution from the

Table 3.1. *Participating codes in the inversion comparison*

Algorithm label	Method	Who
AbelFlex512/1024	Fourier–Hankel	Whitaker/Parker/Eppink
LGFFT	Fourier–Hankel	Janssen/Stolte/Roeterdink
Iterative inversion	Iterative parameter optimization	Vrakking
BASEX	Basis-set expansion	Reisler/Dribinski
Onion	Onion peeling	Loock/Manzhos

expansion coefficients using the equation:

$$P(v) = \frac{1}{(P(v))_{\max}} \sum_{k=0}^{K_x-1} \sum_{m=0}^{K_z-1} C_{km} b_{km} \left(\frac{e}{k^2+m^2} \right)^{k^2+m^2} v^{2(k^2+m^2+1)} e^{-v^2}, \quad (3.43)$$

with

$$b_{km} = \left[\frac{(k^2+m^2)^{k^2+m^2}}{(k^2)^{k^2}(m^2)^{m^2}} \right] \int_{-1}^{1} (1-\tau^2)^{k^2} (\tau^2)^{m^2} d\tau. \quad (3.44)$$

3.7 Algorithm evaluation

In this section a comparison is presented between five different inversion algorithms that we could use thanks to the cooperation of their developers (Table 3.1). First a short description of each program is given, followed by some test criteria and test results. A series of 17 simulated images and three experimental images have been processed using each inversion program. The reconstructed velocity distributions were analysed using a common routine for obtaining the speed and angular profiles. This excludes differences in the analysis procedures after the actual image inversion, which might otherwise obviate a fair comparison.

3.7.1 Algorithms

The AbelFlex code (available from KUN, Nijmegen) uses the Fourier–Hankel inversion method, with additional features for selectively centring, smoothing, and symmetrizing the raw image prior to inversion. The source code was taken from Abel5 of J. Myers (jd_myers@pnl.gov). The Abel5 code (1993) itself included modified versions of codes originally written by B.J. Whitaker ('Hankel') and D. Huestis ('Abel'). Parker and Eppink found the Hankel code to perform better at the time, i.e. produce sharper speed distributions, and so selected that one for AbelFlex. The calculation is performed either on a basis of 512^2 pixels or 1024^2 pixels. Speed and angular distributions can be produced either automatically or manually. The C++ code is compiled for Windows and yields a calculation time of 32 s for a centred 512×512 image on a PC with a 1 GHz Celeron processor.

In the Large Grid FFT (LGFFT) code the image width is extended up to 2048^2 or 4096^2 pixels, possibly padded with the average background value of the data taken from the first 1–3 lines of the image. This removes additional step functions that potentially create artificial features in the inverted image. The enlarged grid improves the quality of the reconstruction compared to the situation where the ion image size is comparable to the inversion grid size [16]. For the present comparison, test images were inverted on a grid of 2096 pixels. The Fortran code, running on a PC, takes about 20 s for 2048×512 or 67 s for 4096×512 pixel images. The code was originally written by M. Janssen, and has been adapted for continuous distributions by W. Roeterdink (VU Amsterdam).

The iterative inversion program of M. Vrakking (AMOLF, Amsterdam) also provides several parameters to play with for finding the centre position and setting the iteration parameters c_1 and c_2. Optionally, the first step can be chosen to be a Fourier–Hankel inversion, or the default of setting the 3-D radial and angular distributions equal to the ones found directly from the experimental image. For each run, the parameters are set in an ASCII file. Several output files are produced, among which the inverted image in Cartesian and polar coordinates, a simulated image which can be compared to the experimental (target) image, and information on the convergence as a function of iteration number. From the reconstructed image in polar coordinates it is easy to obtain the radial distribution directly, but this is not equal to the speed distribution due to the weighting factor of $r \sin \vartheta$. Typically it takes the compiled Fortran code about 3 min on a PC with 950 MHz Pentium processor to perform ten iterations on a 300×300 grid.

The BASEX program of H. Reisler, V. Dribinski and coworkers (USC, Los Angeles) runs under Windows. The program assumes that the image has been centred and has odd dimensions. Some options for symmetrization, centre-line interpolation, regularization, choice of basis-set width, and settings for the output speed and anisotropy distributions (at subpixel resolution) are available. Typically an inversion takes about 40 s for a 501×501 image on a PC with a 1 GHz Celeron processor.

The Onion peeling program of H. Loock and S. Manzhos (Queens University, Canada) runs under Windows, and provides options for geometric corrections such as squeeze and tilt factors, and produces on-screen output of velocity and angular profiles after the inversion. Angular fitting of up to six bipolar moments (using Legendre polynomials) is possible through the graphical user interface. Optionally, back-simulated images can be produced on the basis of the velocity and angular profiles. A typical inversion of a 500×500 sample image takes 40 s on a PC with 1 GHz Celeron processor. We note that the results presented in this chapter suffered from a conversion from floating point to integer arithmetic that has since been corrected. As a result the inversion times reported can probably be improved upon.

3.7.2 Test criteria and comparison method

In making the comparison between the different algorithms, we need to keep in mind what dynamic information can be extracted from charged particle images. First of all, the radial information in the reconstructed image must be transformed into a 3-D speed distribution, which is directly connected to the kinetic energy distribution to be probed. Each ring in the experimental image corresponds to a Newton sphere in velocity space, and thus a different dynamic production channel. The kinetic energy resolution attainable is thus directly related to the radial resolution in the recovered image. Branching ratios between the various product channels represent another important dynamic observable. These can be extracted from the area under each peak in the speed distribution, so distortion of the recovered speed distribution is an issue. The speed peak positions in the recovered distribution also need to be at the right place, in order to obtain reliable values of the kinetic energies. The 3-D angular distribution of each channel should be reproduced well. The spatial anisotropy is a particularly important function to be probed in many dynamics studies. The method used in this comparison for extracting speed and angular distributions from reconstructed images is described in the Appendix to this chapter.

Now it is clear what information we want to extract, let's take a look at the problems that occur to obscure this information from an experimentalist viewpoint. Clearly, the issue of signal-to-noise ratio is an important one. Experimental images suffer from *signal noise* (each single intensity measurement suffers from Poisson statistics), *background noise* (from dark current counts and CCD read-out noise), but also noise from interfering particles that are not of interest for the particular study, which we may denote as *underground noise*.

As described in Chapter 2, the use of *event counting* can help to alleviate signal noise, detector inhomogeneities, and minimize background noise, but as a result the experimental images can have large point-to-point fluctuations, and a question arises as to how many events are enough for a reliable statistical determination of the dynamic information. The application of velocity map imaging has circumvented the problem of blurred images due to the size of the ionization volume. The process of mapping is achieved with electrostatic lenses, which generally perform quite well but are never free from residual aberrations. These are translated into slight distortions of the image. Especially with electron imaging, stray magnetic

Table 3.2. *Overview of used sample projection images. The letter 'c' in the filename denotes clean rings, whereas 'n' denotes 100% signal noise*

No.	Filename	size (pix)	ΔR(pix)	Remarks
1	cMany/nMany	500×500	2	21 rings cover full chip, $\beta = 2,0,-1$
2	cLess/nLess	500×500	2	9 rings, $\beta = 2,-1$
3	cFew/nFew	500×500	2	5 rings, $\beta = 2,0,-1$
4	cFewDI	500×500	2	like #3, differing intensities
5	cNoOuter	500×500	2	15 rings towards centre of image
6	cSmall	500×500	2	8 rings with small radii ($R_0 = 5,10,15,..$)
7	BcLess/BnLess	500×500	2	like #2, with background noise
8	UcLess/UnLess	500×500	2	like #2, with underground noise
9	cLess_dist	500×500	2	like #2, distorted
10	cFew_dist	500×500	2	like #3, distorted
11	cLess_w1	250×250	1	like #2, 2×2 compressed
12	cLess_w05	125×125	0.5	like # 2, 4 × 4 compressed
13	Exp1	760×760		Event counted, 6.9×10^6 events
14	Exp2	380×380		like #13, 2 × 2 compressed
15	Exp3	380×380		1.8×10^5 events, noisier image

and electric fields can also cause similar imperfections, even with shielding precautions.

In order to address these issues, we used a set of 17 simulated images and three experimental (event-counted) images on all algorithms. Table 3.2 gives an overview of the sample images.

For the synthetic images (file 1–12), the output of each program can be directly compared to a 'perfect slice' through the speed distribution, which is also produced by the simulation program. We assumed for each ring the following 3-D velocity distribution:

$$I(r, \vartheta) = f(r) \times g(\vartheta),$$

with

$$f(r) = \frac{I_0}{4\pi R_0^2} \exp\left[-\left(\frac{r - R_0}{\Delta R}\right)^2\right]$$

and

$$g(\vartheta) = \frac{1}{4\pi}[1 + \beta(3\cos^2\vartheta - 1)/2].$$

Because the parameters ($I_0, R_0, \Delta R, \beta$) are known, the reconstructed values of these parameters can be compared to the input values. From the 3-D distribution $I(r, \theta)$, 2-D 'crush' and 'slice' images are produced, which correspond to the

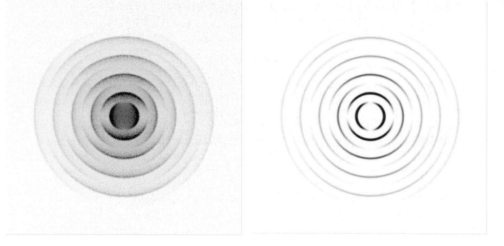

Fig. 3.6. Example of a simulated image ('BnLess', Table 3.2) used in the comparative tests of the reconstruction alorithms (a): projected data ('crush' image); (b): vertical cut through the 3-D distribution ('slice' image).

projected data and ideal reconstruction, respectively. Sample images are produced by summations of crush and slice images that contain single rings. Signal noise, background noise, underground noise and distortions are added afterwards. The distortion applied is a slight skewing of the image along the $y = x$ axis (taking x, y as the detector plane) in order to simulate non-roundness.

In Fig. 3.6 an example simulated image ('BnLess' in Table 3.2) is shown, with the crush data left and the slice data right. It displays nine rings with pure perpendicular or parallel angular character, and with added background and signal noise to the crush data. Figure 3.7 shows the same slice image in polar coordinates, obtained according to the procedure described in the Appendix. The data shown in Fig. 3.2 described earlier corresponds to the data set 'Expt2' in Table 3.2. This sample image is particularly suitable for the inversion comparison, since each inversion method is required to recover the complex ring structure well.

3.8 Test results

Because there is no space here to present results on all the sample images, we select a few which we found most representative.

3.8.1 Clean images

The first test is to see if the inversion programs perform well with input simulated images without added noise or distortions (clean images). This is a simple check that

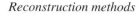

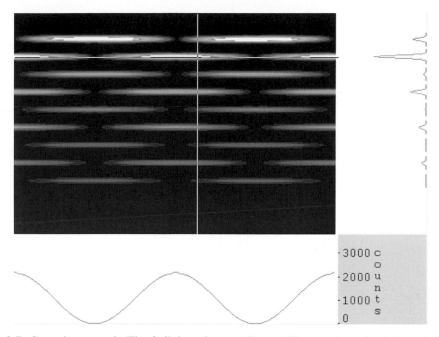

Fig. 3.7. Same image as in Fig. 3.6b in polar coordinates. The panels to the right and at the bottom show cuts through the data taken on the vertical and horizontal lines drawn through the image.

the inversion programs don't have any systematic problems or coding errors, and we expected all programs to perform very well in this test. In Fig. 3.8 speed distributions from the various inversion methods are displayed for the sample image 'cLess'. The label 'produced' denotes that the corresponding trace has been produced by the inversion program itself; else by the speed extraction routine described in the Appendix. The general appearance of the speed distributions (reported in pixel units) look very much alike. However, the AbelFlex and LGFFT methods (both Fourier–Hankel) introduce noise at the top end of the speed distribution. Upon inspection of the corresponding images, this noise appears along the horizontal (x) axis and seems to come from interference effects in the FFT routines.

In order to look at this more closely, we characterized the peaks according to their position, width, height and area. These parameters are important for the correct recovery of key dynamical quantities such as the absolute kinetic energy of a particular channel, the energy resolution and branching ratios. For this purpose, each speed peak was fitted to a Gaussian of the form $y = y_0 + \frac{A}{w\sqrt{\pi/2}} \exp[\frac{-2(x-\mu)^2}{w^2}]$, where μ, w, A represent the peak position, width and area.

Figures 3.9–3.14 show calculations from sample image 'cFew'. In Fig. 3.9, the peak centre deviation is shown. Because the simulated images are not 3-D

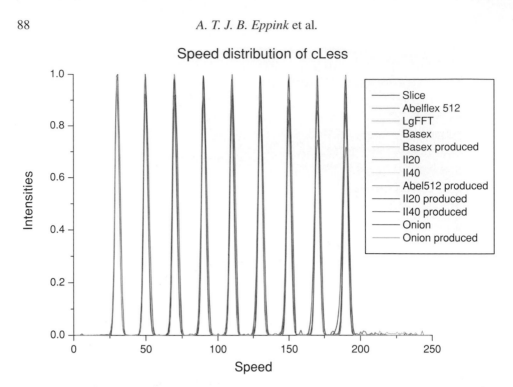

Fig. 3.8. Speed distributions from a clean sample image ('cLess'). This image is similar to Fig. 3.6 but without noise. The figure show comparative results for the various inversion methods tested. The trace labelled 'Slice' represents the true (input) speed distribution. II20 and II40 denote data inversion by Vrakking's iterative algorithm [13] with 20 and 40 iterations, respectively. The label 'produced' means that the corresponding trace has been produced using the internal routines of the relevant inversion program, otherwise the method described in the Appendix has been employed.

normalized (we use the 3-D *radial* function $f(r)$ instead of a normalized *speed* function), all traces show decreasing deviations with increasing speed. Because of this the best comparison is done with the trace obtained from the 'slice'. Here, we see that almost perfect overlap is found for AbelFlex512, LGFFT, BASEX and II40. The Onion method shifts all peaks by 0.2 pixels toward the centre of the image. We also found that method II40 yielded better results than II20, and we will only report results using the higher iteration method from now on.

Figure 3.10 shows the peak widths obtained for all rings in 'cFew'. We see that AbelFlex512 and BASEX come closest to recovering the true slice (within 0.02 pixels), although LGFFT is comparable. The Onion method produces slightly wider peaks (max deviation 0.1 pixels), whilst II40 produces the highest deviations of 0.35 pixels overestimation. Figure 3.11 shows the peak heights. Considering the narrowness of the peaks used, the values are reproduced very well for all methods, with a slight preference for BASEX, and least for II40. Figure 3.12 shows the peak areas, which tests the combination of each method's ability to recover peak

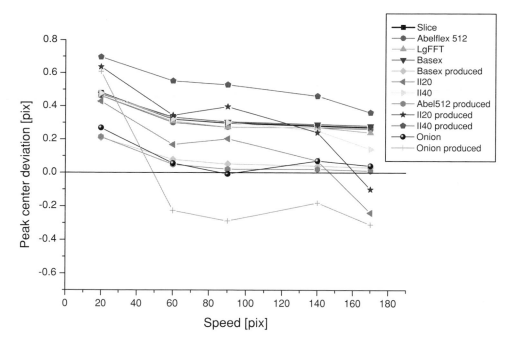

Fig. 3.9. Recovered speed peak centre deviations from the true positions in pixel units for the various inversion methods obtained using the input image 'cFew' (see Table 3.2).

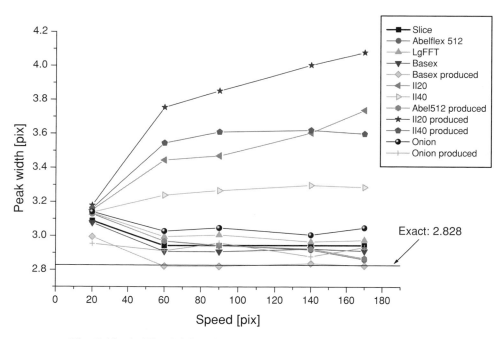

Fig. 3.10. As Fig. 3.9 but showing the recovered speed peak widths.

A. T. J. B. Eppink et al.

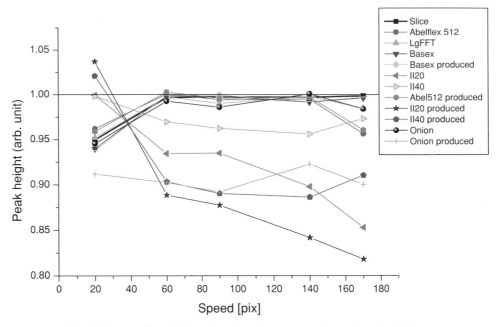

Fig. 3.11. As Fig. 3.9 but showing the recovered speed peak heights.

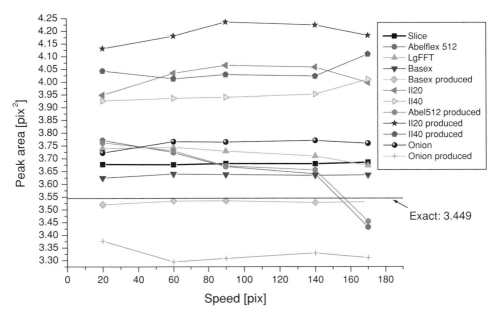

Fig. 3.12. As Fig. 3.9 but showing the recovered speed peak areas.

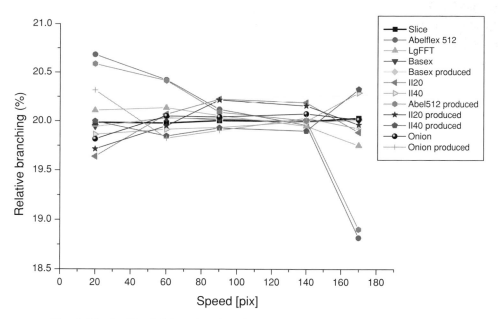

Fig. 3.13. As Fig. 3.9 but showing the recovered relative branching ratios.

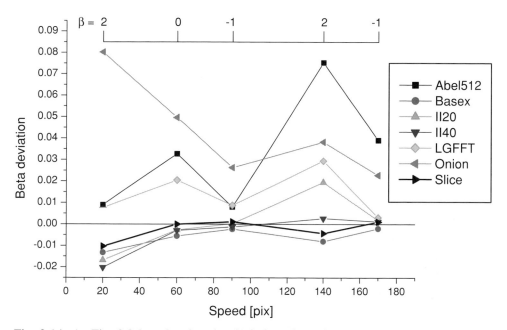

Fig. 3.14. As Fig. 3.9 but showing the deviations from the true value for the recovered anisotropy parameters (β's).

width and height. Now the results from BASEX lie closest to the input data, then LGFFT, AbelFlex512, Onion and II40. As this property is connected to the important branching ratios, we show this separately in Fig. 3.13. Now the best methods are BASEX, Onion and II40, whereas LGFFT and AbelFlex512 do worse.

In Fig. 3.14 the angular distributions of each peak have been fitted and compared to the input value of the beta parameter. Overall, the beta values are reproduced to high accuracy by each program. The best results come from BASEX (where the maximum deviation from the input data is 0.005!) and II40 (0.01), whereas LGFFT (0.02), AbelFlex512 (0.08) and Onion (0.09) show a little less accuracy. The last three methods also produce the greatest centreline noise, which adds spurious counts at $\vartheta = 0°$ and $180°$.

3.8.2 Influence of noise

Now we have seen that all inversion programs behave quite well on simulated images without noise, let's see what happens if we add noise to them. In Fig. 3.15 six reconstructed images of the 'UnLess' simulated data set are shown, that is, a data set with both signal noise and underground noise added. In order to show where the noise is amplified in the reconstruction, the intensity of each reconstructed image

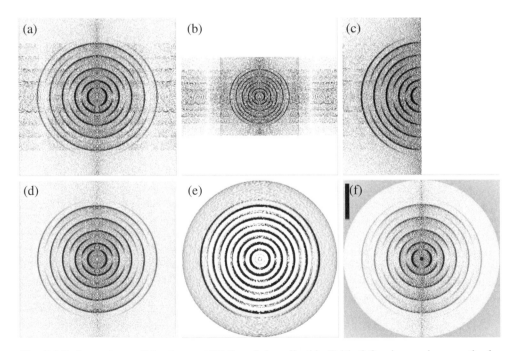

Fig. 3.15. Inversions of the data set 'UnLess' described in Table 3.2 using various methods. Upper row: (a) AbelFlex512, (b) AbelFlex1024, (c) LGFFT. Second row: (d) BASEX, (e) II40, and (f) Onion peeling.

has been expanded so that the recovered rings are saturated. The Fourier–Hankel methods (Fig. 3.15a–c) show noise on the centreline axis, but also on specific lines where the noise appears across the image. The AbelFlex1024 (Fig. 3.15b) method shows artificial effects due to the step function introduced by adding width to the image without padding it with background noise. In the LGFFT method (Fig. 3.15c) – which yields a half output image – the padding has been done with an average value of the background, and as a result less noise is seen in the outer regions of the recovered image. The Onion method (Fig. 3.15f) also shows strong centreline noise, due to the error propagation toward the centre in the peeling process. The centre dot appears because the program cuts out the centre of the image. BASEX (Fig. 3.15d) shows much less centreline noise, and no noise amplification on the outside of the image. The II40 (Fig. 3.15e) method also shows no centreline noise, but the intensities between the rings are artificially clipped to zero instead of having the proper underground noise value. This image has been blown up more strongly than the others to show the artificial effects close to the centre of the image, which in different cases can appear even more strongly. The noise towards the outside of Fig. 3.15e therefore appears stronger than in the other images, but is in fact comparable to the other images and close to the input underground noise value.

In Fig. 3.16 recovered speed distributions are plotted for the 'BcLess' data set, that is, the same image as 'cLess' but with added background noise. Because of the

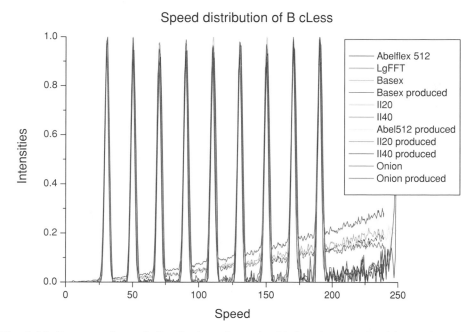

Fig. 3.16. Recovered speed distributions from the 'BcLess' synthesized image in which background noise was added to the clean image 'cLess'.

weighing factor, $R \sin \vartheta$, in the speed distribution (see the Appendix), the centre-line noise appears much weaker and the 'outside' noise much stronger in the speed distributions. We can see clearly that the baseline is lifted towards the outside of the reconstructed speed distribution. The effect is strongest for the Onion method. BASEX, AbelFlex512 and LGFFT show the effect more weakly and are more or less comparable. Only II40 has no baseline, but also on inspection of the inverted image, the intensities between the rings are clipped to zero. We suspect an inversion correction parameter setting to be responsible for the 'over correction' in the inversion procedure. Inversions on the 'UcLess' data set (which includes added underground noise) are presented in Fig. 3.17, and again the II40 method clipped the recovered signal to zero, which supports this assumption. The results for the other methods on the 'UcLess' data are comparable with those from 'cLess'; the Onion peeling algorithm returns the most underground noise, followed by BASEX, LGFFT, AbelFlex512 and II40. In all cases, the speed distributions produced from BASEX method using its internal procedure seem superior.

An interestingly different behaviour occurs if we add signal noise to the image. Figure 3.18 shows the results for the 'nLess' data set. Only II40 shows sharp peaks, but we see again the effect of clipping to zero between the peaks. All the other

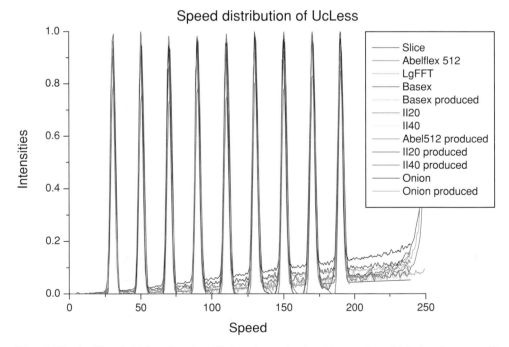

Fig. 3.17. As Fig. 3.16 but for the 'UcLess' synthesized image in which 'underground' noise was added to the image.

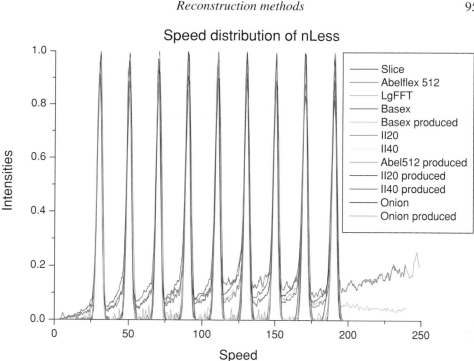

Fig. 3.18. As Fig. 3.16 but for the 'nLess' synthesized image in which signal noise was added to the image.

methods show tails in the recovered speed peaks toward smaller radii. The BASEX and Onion peeling programs produce better speed distributions using their internal routines, although on close inspection we see that the inverted images suffer from the same 'tail' effect albeit to a lesser degree.

Figure 3.19 shows the difference between the recovered beta parameters and the input values for the speed peaks obtained from the 'BnLess' data (in which signal noise and background noise are added to the image). The trend here shows alternating behaviour for the parallel and perpendicular channels, with the highest decrease of anisotropy for the perpendicular channels. Il40 is the best inversion method, with a maximum deviation between the input and recovered value of 0.085, followed by BASEX and LGFFT (0.13 maximum deviation). Worst are the Onion peeling and AbelFlex512 methods (0.23 maximum deviation). The decrease in the recovered anisotropy due to the addition of noise is expected to be strongest for extracted velocity distributions exhibiting the most centreline noise (i.e. the Fourier–Hankel methods). The reason that the iterative inversion method is the superior in this test is probably because it suffers the least from this problem.

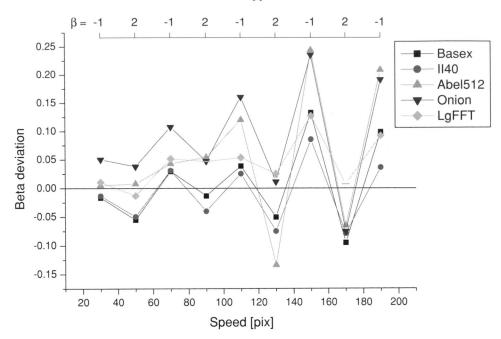

Fig. 3.19. Deviations from the true anisotropy parameter recovered from the 'BnLess' synthesized image in which signal noise and background noise were added to the image.

3.8.3 Influence of distortion

In Fig. 3.20a the input image 'cLess' is shown in polar coordinates, with the angle θ along the horizontal axis and radius R along the vertical axis. Skewing of the image along the $y = x$ axis is apparent as 'wiggles' in the horizontal stripes (which correspond to the rings in the original image in Cartesian coordinates). These wiggles are absent in the AbelFlex512, LGFFT and BASEX inversions (panel b shows the results for AbelFlex512). This is due to an implicit symmetrization of the reconstructed images by these methods. The II40 and Onion peeling methods, however, (panel c and d) preserve the original distortion in the inversion. We note that the implementation of the Onion peeling method written by Manzhos and Looke [12] allows the possibility of correcting for vertical elliptic distortions in the raw and inverted images. This option has not been used for this test.

In Fig. 3.21 the speed peak deviations (a), widths (b), heights (c) and areas (d) are shown for this image. The overall trend is that the width increases and the height decreases as a function of speed, but these effects cancel out when the area is determined. This is an important aspect when obtaining branching ratios from images. The Onion peeling program performs best in comparisons of the recovered to input peak widths and heights, but BASEX scores best on the peak areas. From the point of view of kinetic energy resolution, AbelFlex512 produces

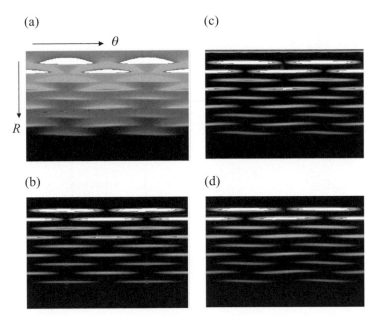

Fig. 3.20. Images in polar coordinates. (a) Input image 'cLess_dist', in which vertical elliptic distortion has been added to the clean image 'cLess'. Inverted images obtained using the methods: (b) AbelFlex512, (c) II40 and (d) Onion peeling.

the sharpest and most faithful peaks. All the results obtained on the distorted image data are comparable to those obtained from undistorted data (Figs. 3.9–3.14), from which we conclude that a slight image distortion does not cause dramatic effects.

In Fig. 3.22, the deviations from the input beta parameter are shown for the distorted data set. The effects of the distortion on the recovered β values are more extreme than those characterizing the recovered speed distributions. In this test the BASEX method performed best, followed by LGFFT, Onion and II40. AbelFlex512 deviates from the true values the most here.

3.8.4 Reconstruction accuracy in numbers

We may now summarize the results found for peak position deviations (a measure for the kinetic energy accuracy), relative branching ratios and beta parameter deviations, and how noise and distortion affect the numbers.

In Table 3.3 the mean speed shift (averaged over all peaks) is shown for the various methods on three sample images: without noise ('cFew'), with distortion ('cLess_dist') and with background and signal noise ('BnLess'). Using our speed routine (the nonitalic rows), we see that all methods are very good at recovering the speed distribution, except for II20 and the Onion peeling methods which

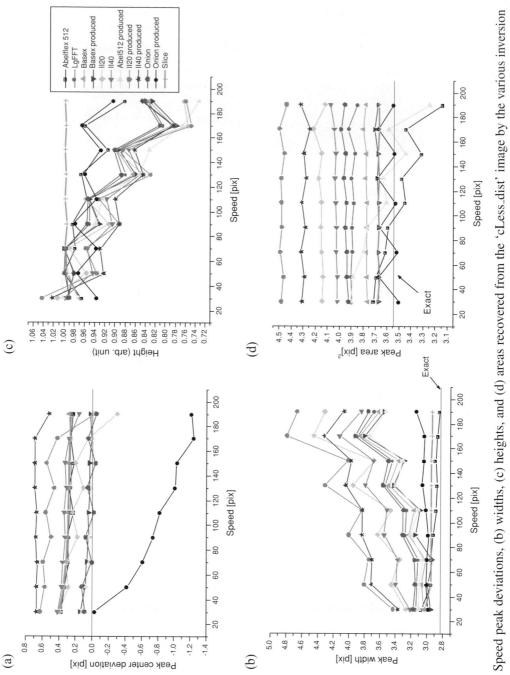

Fig. 3.21. (a) Speed peak deviations, (b) widths, (c) heights, and (d) areas recovered from the 'cLess.dist' image by the various inversion methods.

Table 3.3. *Mean speed shift*

Method	cFew		cLess_dist		BnLess	
Abelflex	−0.017	±0.007	0.010	±0.021	0.012	±0.018
Abel Prod.	*−0.267*	*±0.008*	*0.193*	*±0.185*	*0.231*	*±0.086*
LGFFT	−0.019	±0.008	0.031	±0.029	0.038	±0.020
Basex	0.006	±0.008	−0.221	±0.030	−0.142	±0.042
Basex Prod.	*−0.248*	*±0.012*	*−0.115*	*±0.178*	*0.017*	*±0.064*
II20	−0.204	±0.181	0.042	±0.054	0.240	±0.055
II20 Prod.	*−0.026*	*±0.206*	*0.383*	*±0.061*	*0.499*	*±0.056*
II40	−0.031	±0.057	−0.250	±0.009	−0.245	±0.009
II40 Prod.	*0.190*	*±0.060*	*−0.243*	*±0.031*	*−0.310*	*±0.028*
Onion	−0.260	±0.049	−1.068	±0.351	−0.391	±0.094
Onion Prod.	*−0.409*	*±0.305*	*0.025*	*±0.024*	*0.119*	*±0.032*

Unit: pixel.

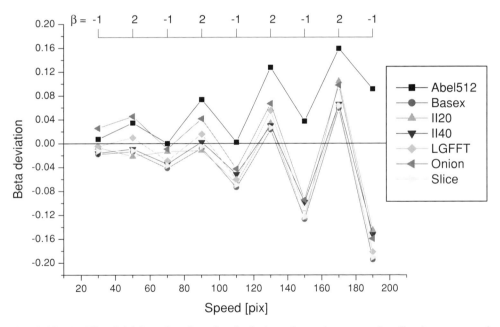

Fig. 3.22. As Fig. 3.21 but showing the deviations from the true value for the recovered beta parameters.

are a little bit less accurate. Adding distortion, the best results are obtained from AbelFlex and LGFFT, more shift is seen for BASEX and II40, and the Onion peeling method shows a mean shift of one pixel. The same behaviour is observed when noise is added to the image, where the Fourier–Hankel methods are the most reliable.

better, with BASEX itself producing the best speed distributions from the viewpoint of baseline shift, signal-to-noise and peak sharpness. The main advantage of the BASEX method over the iterative inversion is its speed and relative ease of use.

With respect to the reproduction of speed and angular distributions, we conclude that all programs perform satisfactorily. A mean speed shift of less than 0.25 pixel is generally achievable. Branching ratios are also reproduced well: deviations of about 5% are achievable for BASEX, LGFFT and Il40; about 10% for AbelFlex and Onion peeling. We have also shown that beta parameters are determinable to very high accuracies ($< \pm 0.01$) by all methods. In the case of very noisy images, the iterative inversion method might be preferable, due to the absence of centreline noise. It is noteworthy that higher accuracy in the branching ratios was generally achieved when the speed distributions were determined using the internal routines in the codes tested rather than the common routine we used for the comparative tests. This was particularly true for the BASEX and Onion peeling codes. In the case of BASEX, this is due to the exact speed formulation from the expansion parameters, and in the case of Onion peeling a round-off failure from floating point to integer arithmetic is possibly responsible for the worse results presented here.

We would like to point out that whatever the inversion method careful attention needs to be paid to proper centring of the input image. In some methods care must also be taken in setting the inversion parameters. For BASEX this means a proper choice of the basis set (form and width) and for the iterative inversion the choice iteration parameters c_1 and c_2. Also experimental artefacts can make one method preferable above others. Where in this comparison one might conclude that the BASEX and iterative inversion methods are the best, for other images the Onion peeling method might be preferable. This would occur, for example, for an image with an intense background feature at the centre or along a laser axis cutting through the ring structure. For all methods, correcting for image distortions prior to the inversion is a good idea – as partially incorporated in the Onion peeling program.

3.9 Forward convolution

The measurement of differential cross-sections for rotational inelastic scattering and reactive bimolecular collisions could, in principle, be made by performing an imaging experiment and the applying any of the data recovery techniques just described. However, experimental considerations make this impracticable.

It is currently impossible to make molecular beam pulses much shorter than 10 µs in duration; indeed, in most experiments reported to date the pulses are typically several 100 µs long. Thus the two molecular beams interact with (scatter from) each other for a time that is much longer than the laser pulse which state selectively ionizes the product species that is to be imaged. As a result product particles scattered at low velocity in the laboratory frame accumulate in the detection volume at the expense of particles scattered with a fast velocity in the laboratory frame, which escape from the detection volume. When the laser pulse interrogates the scattering centre, typically on the nanosecond time scale, the ions that are produced are consequently weighted heavily in favour of those scattered with a small velocity in the laboratory frame. Because the velocity measured in the laboratory frame is the vector sum of the centre-of-mass and the scattered velocity,

$$\mathbf{v}_{\text{lab}} = \mathbf{v}_{\text{COM}} + \mathbf{v}_{\text{scat}}, \tag{3.45}$$

the upshot is that the measured image is not cylindrically symmetric with respect to $\mathbf{v}_{\text{lab}}$, and none of the methods hitherto described are valid ways of recovering $\mathbf{v}_{\text{scat}}$.

A number of approaches to this problem have been proposed [17,18]. They are all based on the idea of guessing a differential cross-section and then simulating the image by *forward convolution*. The calculated image is then compared with the observed image and the process iterated by refining the input differential cross-section until reasonable agreement is found between the simulated and observed images. The difference between the approaches is the way in which one decides on what is meant by 'reasonable agreement'.

We can simulate the expected image in the following way: we express the velocity mapped ion image across the image plane (x, z) as:

$$I(v_x, v_z) = P_{2D} \left\{ \int_{-\infty}^{o} dt \int_{\substack{\text{detection} \\ \text{volume}}} d\mathbf{R}' \iint_{V_1, V_2} dv_1 dv_2 \right.$$
$$\left. \times n_1(\mathbf{R}, v_1, t) n_2(\mathbf{R}, v_2, t) f(\mathbf{R}') \sigma(\vartheta) \right\}, \tag{3.46}$$

where $t = 0$ is the time at which the ionizing laser is fired, $\mathbf{R}$ and $\mathbf{R}'$ describe the position vectors of the particles before and after ionization respectively, $n_1(\mathbf{R}, v_1, t)$ and $n_2(\mathbf{R}, v_2, t)$ are the number densities in the two molecular beams as a function of position, velocity component, v_i, and time respectively, $f(\mathbf{R}')$ is the ionization efficiency and $\sigma(\vartheta)$ is the differential cross-section that we wish to determine. The operator P_{2D} symbolizes the projection of the 3-D velocity distribution of scattered products onto the detector surface analogous to equation (3.16).

The functions $n_1(\mathbf{R}, v_1, t)$ and $n_2(\mathbf{R}, v_2, t)$ can be characterized by experiment. The particle density throughout each of the molecular beam pulses can be obtained by directly ionizing a component in the beam, or a convenient tracer molecule such as NO seeded in the beam, and measuring the total ion yield as a function of the time delay between opening the molecular beam valve and firing the ionization

Amsterdam), Marc Vrakking (AMOLF, Amsterdam), Hans-Peter Loock and Sergei Manzhos (Queens University, Kingston) for sharing their algorithms and preprints with us during the course of writing this contribution. This has allowed us to perform a comparative test of the various inversion algorithms for the first time. We would also like to thank Hiroshi Kohguchi (IMS, Okazaki) for discussions concerning the forward convolution methods described, and Eckart Wrede (Durham) and Mike Ashfold (University of Bristol) for supplying Fig. 3.1. Special thanks go to Eleni Niktari (FORTH, Heraklion) for helping with the comparisons and to David Parker (Nijmegen) for discussions and suggestions.

References

1. A. J. R. Heck, D. W. Chandler, *Annu. Rev. Phys. Chem.* **46**, 335 (1995); B. J. Whitaker. In *Research in Chemical Kinetics*, eds. R. G. Compton, G. Hancock, (Elsevier, Amsterdam, 1993), vol. 1, p. 307.
2. E. Wrede, E. R. Wouters, M. Beckert, R. N. Dixon, M. N. R. Ashfold, *J. Chem. Phys.* **116**, 6064 (2002).
3. D. H. Parker, A. Eppink, *J. Chem. Phys.* **107**, 2357 (1997).
4. C. R. Dasch, *App. Opt.* **31**, 1146 (1992).
5. R. N. Bracewell, *The Fourier Transform and its Applications*. (McGraw-Hill, New York, 1978).
6. L. M. Smith, D. R. Keefer, *J. Quant. Spectrosc. Radiat. Transfer* **39**, 367 (1988).
7. W. Castleman, *Digital Image Processing*. (Prentice-Hall, London, 1979).
8. W. H. Press, B. P. Flannery, S. A. Teukolsky, W. T. Vetterling, *Numerical Recipes in C*. (Cambridge University Press, Cambridge, 1988).
9. S. M. Candel, *Comput. Phys. Comm.* **23**, 343 (1981).
10. C. Bordas, F. Paulig, H. Helm, D. L. Huestis, *Rev. Sci. Inst.* **67**, 2257 (1996).
11. J. Winterhalter, D. Maier, J. Hornerkamp, V. Schyja, H. Helm, *J. Chem. Phys.* **110**, 11187 (1999).
12. S. Manzhos, H.-P. Loock, *Comp. Phys. Commun.*, in press (2002).
13. M. J. J. Vrakking, *Rev. Sci. Instrum.* **72**, 4084 (2001).
14. V. Dribinski, A. Ossadtchi, V. Mandelshtam, H. Reisler, *Rev. Sci. Instrum.*, in press (2002).
15. A. N. Tikhonov, *Soviet Math. Dokl.* **4**, 1035 (1963).
16. W. Roeterdink, Ph.D. Thesis, Free University of Amsterdam, 2002.
17. L. S. Bontuyan, A. G. Suits, P. L. Houston, B. J. Whitaker, *J. Phys. Chem.* **97**, 6342 (1993); A. G. Suits, L. S. Bontuyan, P. L. Houston, B. J. Whitaker, *J. Chem. Phys.* **96**, 8618 (1992); N. Yonekura, C. Gebauer, H. Kohguchi, T. Suzuki, *Rev. Sci. Instrum.* **70**, 3265 (1999); K. T. Lorenz, M. S. Westley, D. W. Chandler, *Phys. Chem. Chemical Physics* **2**, 481 (2000); M. S. Westley, K. T. Lorenz, D. W. Chandler, P. L. Houston, *J. Chem. Phys.* **114**, 2669 (2001).
18. G. C. McBane, Imsim (http://www.chemistry.ohio-state.edu/~mcbane/research.html, 2000).

4

Orientation and alignment

T. PETER RAKITZIS

4.1 Introduction

Several vectors play an important role in the dynamics of molecular photodissociation. Well known examples include the angular momentum of the parent molecule $\mathbf{J}_i$, the transition dipole moment $\boldsymbol{\mu}$ of the dissociating transition, the relative recoil velocity $\mathbf{v}$, and the angular momentum of the fragments $\mathbf{J}_f$.

In previous chapters we have introduced the idea that the measurement of correlations between these vectors can yield deep insight into the dynamics of the dissociation process. For example, we have seen that for prompt photodissociation of a small molecule the correlation between $\boldsymbol{\mu}$ and $\mathbf{v}$ (or the wavevector $\mathbf{k}'$) is determined by the relative symmetry of the ground and excited states, and that the angular distribution $I(\theta)$ of the velocity $\mathbf{v}$ of the photofragments about the photolysis polarization is given by the, by now familiar, expression [1]:

$$I(\vartheta) \propto 1 + \beta \, P_2(\cos \vartheta), \tag{4.1}$$

where ϑ is the angle between the photolysis laser polarization vector and $\mathbf{v}$. The anisotropy parameter β ranges from $+2$ (for a parallel transition, for which $\boldsymbol{\mu}$ and $\mathbf{v}$ are parallel) to -1 (for a perpendicular transition, for which $\boldsymbol{\mu}$ and $\mathbf{v}$ are perpendicular). Nonlimiting values of β result (a) for bent polyatomic molecules (e.g. NO_2) for which $\boldsymbol{\mu}$ is neither exactly parallel nor perpendicular to $\mathbf{v}$, (b) for coherent excitation of parallel and perpendicular transitions and (c) for molecules with dissociation lifetimes comparable to a rotational period, or any combination of (a), (b) and (c). Therefore, measurements of the anisotropy parameter β can only be interpreted in the light of previous knowledge about the excited state(s) of the parent molecule. Furthermore because β only measures the correlation between $\boldsymbol{\mu}$ and the final relative velocity $\mathbf{v}$ of the photofragments it is insensitive to the overall shape of the excited state potential energy surface. In short, many details of dissociation dynamics cannot be determined, except for the very simplest cases, by measuring a single number, the β parameter.

By contrast, the correlations between other observables associated with the photofragment angular momentum distributions are sensitive to many more details of the dissociation process. A complete quantum mechanical treatment of photofragment polarization has been given by Siebbeles *et al.* [2], and a similar formalism will be used here [3]. The aim of this chapter is to examine what the photofragment angular momentum distributions and vector correlations can tell us about the photodissociation process. Initially, the treatment presented will refer to the prompt photodissociation of diatomic molecules. The description of the angular momentum distribution in the molecular frame will follow the $a_q^{(k)}(p)$ parameter formalism [3], and for simplicity the index k will be limited to $k \leq 2$ (more complete descriptions are given elsewhere [3, 4]).

4.2 Formalism

For a pure parallel transition, the excited parent molecule is cylindrically symmetric (the bond axis, $\mathbf{v}$, and $\boldsymbol{\mu}$ are parallel; see Fig. 4.1a). The angular momentum of the resulting cylindrically symmetric photofragments can be described by cylindrically symmetric polarization parameters, $a_0^{(2)}(\|)$ ($q = 0$ only), and the photofragment

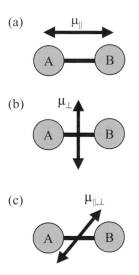

Fig. 4.1. (a) For a pure parallel transition in the axial recoil limit, the bond axis, the transition dipole moment $\boldsymbol{\mu}$, and the recoil direction $\mathbf{v}$ are all parallel, so that the photofragments possess an axis of cylindrical symmetry about these vectors. (b) For a pure perpendicular transition, $\boldsymbol{\mu}$ is perpendicular to the bond axis and $\mathbf{v}$, so that the photofragments possess two orthogonal planes of symmetry. (c) For a mixed transition, where $\boldsymbol{\mu}$ is neither parallel nor perpendicular to the bond axis, the photofragments possess only one plane of symmetry.

detection expression is given by:

$$I_{\parallel}(\Theta) = 1 + s_2 a_0^{(2)}(\parallel)\, C_0^2(\Theta), \tag{4.2}$$

where Θ is the angle between the photolysis laser polarization and $\mathbf{v}$. Notice that (4.2) is identical in form to (4.1) [the $C_q^k(\Theta)$ are the reduced spherical harmonics with $\Phi = 0$, given by $[4\pi/(2k+1)]^{1/2}\, Y_q^k(\Theta, 0)$, and thus $C_0^k(\Theta) = P_k(\cos\Theta)$], so that the parameter $a_0^{(2)}(\parallel)$ has a similar interpretation as β: for positive values of $a_0^{(2)}(\parallel)$ the total angular momentum vector $\mathbf{J_f}$ of the detected fragments is preferentially parallel to $\mathbf{v}$, whereas for negative values of $a_0^{(2)}(\parallel)$ $\mathbf{J_f}$ is preferentially perpendicular to $\mathbf{v}$.

For a pure perpendicular transition, $\boldsymbol{\mu}$ is perpendicular to the recoil direction, so that the excited heteronuclear parent molecule possesses two orthogonal planes of symmetry (see Fig. 4.1b). These symmetry planes forbid the existence of the $q = 1$ parameters. The angular momentum of the resulting photofragments can be described by the polarization parameters, $a_0^{(2)}(\perp)$ and $a_2^{(2)}(\perp)$ ($q = 0$ and 2), and the ionization expression is given by:

$$I_{\perp}(\Theta, \Phi) = 1 + s_2\big[a_0^{(2)}(\perp)C_0^2(\Theta) + 2\, a_2^{(2)}(\perp)C_2^2(\Theta)\cos 2\Phi\big]. \tag{4.3}$$

For a coherent superposition of parallel and perpendicular transitions, $\boldsymbol{\mu}$ is at an angle to the recoil direction (see Fig. 4.1c), and there is a plane of symmetry containing all three vectors (for linearly polarized photolysis light only). This reduced symmetry now allows the presence of the terms $\text{Im}[a_1^{(1)}(\parallel, \perp)]$ and the $\text{Re}[a_1^{(2)}(\parallel, \perp)]$. The contribution of these $q = 1$ terms to the ionization expression is given by $I_{\parallel,\perp}(\Theta, \phi)$:

$$I_{\parallel,\perp}(\Theta, \Phi) = s_1\big\{\text{Im}\big[a_1^{(1)}(\parallel, \perp)\big]C_1^1(\Theta)\sin\Phi\big\}$$
$$+ s_2\big\{\text{Re}\big[a_1^{(2)}(\parallel, \perp)\big]C_1^2(\Theta)\cos\Phi\big\}. \tag{4.4}$$

Notice that the symmetry of linearly polarized photolysis light excludes the terms $a_0^{(1)}(\perp)$, $\text{Re}[a_1^{(1)}(\parallel, \perp)]$, and the $\text{Im}[a_1^{(2)}(\parallel, \perp)]$. These parameters can be detected with circularly polarized photolysis light only.

The total detection expression, I_T, is given by a population-dependent and angle-dependent weighted sum of (4.1)–(4.3):

$$I_T(\theta, \Theta, \Phi) = (1 + \beta)\cos^2\theta\, I_{\parallel}(\Theta) + (1 - \beta/2)\sin^2\theta\, I_{\perp}(\Theta, \Phi)$$
$$+ 2\sin\theta\cos\theta\, I_{\parallel,\perp}(\Theta, \Phi). \tag{4.5}$$

The parallel transition detection expression $I_{\parallel}(\Theta)$ is weighted by the well-known parallel transition angular distribution ($\cos^2\Theta$) and the population weighting of $(1 + \beta)$, which vanishes for $\beta = -1$, as expected. Similarly, the perpendicular transition detection expression $I_{\perp}(\Theta, \Phi)$ is weighted by the well-known perpendicular transition angular distribution ($\sin^2 \Theta$) and the population weighting of $(1 - \beta/2)$, which vanishes for $\beta = +2$, as expected. Finally, the interference term should now have as weighting terms twice the square root of the product of the coefficients of the parallel and perpendicular transition expressions. This should yield $2\sqrt{(1 + \beta)(1 - \beta/2)} \sin \Theta \cos \Theta$; however, as can be seen, the factor of $\sqrt{(1 + \beta)(1 - \beta/2)}$ has been omitted and has been implicitly included in the definition of the $q = 1$ parameters, as described in Ref. [4]. The reason for this choice is that the value of the factor $\sqrt{(1 + \beta)(1 - \beta/2)}$ changes rapidly when the β parameter is close to -1 or $+2$, so that the uncertainty in the factor $\sqrt{(1 + \beta)(1 - \beta/2)}$ is very large for relatively small uncertainties in the β parameter.

After introducing the parameters $A_q^k(p)$ (see also Table 2.1):

$$A_0^k(\text{iso}) = \frac{1}{3}\left[(1 + \beta)\, a_0^{(k)}(\|) + 2(1 - \beta/2)a_0^{(k)}(\perp)\right] \tag{4.6a}$$

$$A_0^k(\text{aniso}) = \frac{2}{3}\left[(1 + \beta)\, a_0^{(k)}(\|) - (1 - \beta/2)a_0^{(k)}(\perp)\right] \tag{4.6b}$$

$$A_1^k = \sqrt{\frac{8}{3}}a_1^{(k)}(\|, \perp) \tag{4.6c}$$

$$A_2^k = \sqrt{\frac{32}{3}}\,(1 - \beta/2)\,a_2^{(k)}(\perp). \tag{4.6d}$$

Equation (4.5) can be rearranged into a particularly symmetric form:

$$I_T(\theta, \Theta, \Phi) = 1 + \beta C_0^2(\theta) + s_1 \,\text{Im}\left[A_1^1\right]\left[C_1^1(\Theta)C_1^1(\theta)\right]\sin \Phi + s_2 A_0^2(\text{iso})C_0^2(\Theta)$$

$$+ s_2\left\{A_0^2(\text{aniso})\left[C_0^2(\Theta)C_0^2(\theta)\right] + \text{Re}\left[A_1^2\right]\left[C_1^2(\Theta)C_1^2(\theta)\right]\cos \Phi\right.$$

$$\left. + A_2^2\left[C_2^2(\Theta)C_2^2(\theta)\right]\cos 2\Phi\right\}. \tag{4.7}$$

Explicit expressions for the sensitivity factors s_2 (and s_4) are given for linearly and circularly polarized light for $(2 + 1)$ REMPI in Ref. [4] whereas, for linearly polarized probe light, the sensitivity factor s_2 is equal to the factor $h(2)$ given by Dixon for $1 + 1$ LIF (and $1 + 1$ REMPI), as described in more detail in Ref. [5].

Equation (4.7) provides a complete detection expression for angular momenta for $k = 2$, as it is of the same form as the bipolar moment formalism of Dixon [5], allowing the expression of the $a_q^{(k)}(p)$ in terms of the bipolar moments [6]. However, (4.7) is a molecular frame expression, and cannot be used directly to describe experimental signals. Methods for obtaining the $a_q^{(k)}(p)$ parameters from experimental signals are described elsewhere. Equation (4.7) can be transformed to the laboratory frame, which has been done for Abel-invertible, ion-imaging geometries [4], slice imaging [7], or it can be used in a Monte Carlo simulation of the experiment [3, 8, 9]. Analytical expressions for the effects of photofragment alignment on laboratory signals have been obtained, for certain geometries, for Doppler profiles [5, 6] and ion images [10]. Complete three-dimensional expressions for arbitrary probe and photolysis geometries (which can then be applied to any method of velocity selection) can be expressed analytically.

4.3 Interpretation

The physical interpretation of the $a_q^{(k)}(p)$ parameters, in the axial recoil limit for diatomic molecules, is straightforward [3]. A brief overview is given here, followed by a description of how this interpretation changes for more complicated situations, such as for bent, polyatomic molecules, and for dissociations with significant lifetimes.

4.3.1 Axial recoil of diatomic molecules

The $a_0^{(k)}(\parallel)$ and $a_0^{(k)}(\perp)$ parameters determine the photofragment m-state distributions (orientation and alignment) with respect to the recoil direction for dissociation by parallel and perpendicular transitions respectively, and explicit equations are given elsewhere [3].

The $a_1^{(k)}(\parallel, \perp)$ parameters arise from the interference of at least two coherently excited states accessed by at least one parallel and one perpendicular transition. When two states interfere, there is a well-defined phase difference $\Delta\phi$ between the asymptotic wavefunctions of the two dissociating states. In this case, the parameters $\mathrm{Im}[a_1^{(k)}(\parallel, \perp)]$ and $\mathrm{Re}[a_1^{(k)}(\parallel, \perp)]$ are proportional to $\sin\Delta\phi$ and $\cos\Delta\phi$ respectively. Thus, the $a_1^{(k)}(\parallel, \perp)$ parameters can provide a sensitive probe of $\Delta\phi$. The measurement of $\Delta\phi$ as a function of the photolysis wavelength is very sensitive to the relative potential energy curves of the interfering dissociative states. This type of measurement constitutes a spectroscopy of dissociative states [11].

The $a_2^{(k)}(\perp)$ parameters arise from perpendicular transitions only ($\Omega' = \Omega \pm 1$). For these transitions, the $\Omega' = \Omega \pm 1$ states are excited coherently in the parent molecule, and this coherence breaks the cylindrical symmetry of the parent molecule about the recoil direction (perpendicular to $\mathbf{v}$, and either parallel or perpendicular to the μ–$\mathbf{v}$ plane). The $a_2^{(k)}(\perp)$ parameters are a measure of how this coherence (and symmetry breaking) is maintained in the photofragment angular momentum.

Recently, the $a_q^{(k)}(p)$ have been expressed in terms of physical quantities such as nonadiabatic transition probabilities and asymptotic phase differences [7,12], and values for the $a_q^{(k)}(p)$ have been calculated for the $F(^2P_{3/2})$ photofragments from the photodissociation of HF and DF.

4.3.2 Non-linear molecules

For molecules with nonlinear excited states, the transition dipole moment μ need no longer be either parallel or perpendicular to $\mathbf{v}$. In such a case, even for a single excited state, none of the $a_q^{(k)}$ (p) parameters are forbidden, and the interpretation of the $a_q^{(k)}$ (p) parameters changes somewhat.

In particular, $a_q^{(k)}$ ($\parallel$, $\perp$) parameters no longer arise necessarily from interference effects. If only one excited state contributes to the dissociation, then interference effects *cannot* be a contributing factor. The $\text{Im}[a_1^{(1)}(\parallel, \perp)]$ orientation parameter can now arise from a classical mechanism, in which the force of dissociation of the bent molecule will constrain the molecular photofragments to have a preferential sense of rotation in the μ–$\mathbf{v}$ plane. Such an example was observed for the NO photofragments from the photodissociation of NO_2 [13]. Furthermore, a correlation of the photofragment angular momentum to μ can appear as a nonzero $\text{Re}[a_1^{(2)}(\parallel, \perp)]$ parameter with respect to $\mathbf{v}$, as seen for the $O(^3P_J)$ photofragments from the photodissociation of NO_2 [14].

For example, consider the photodissociation of a bent triatomic molecule AB_2 with a bond angle of about $150°$ and with μ parallel to the B–B axis (see Fig. 4.2a). Let us assume that the photodissociation produces photofragments B with angular momentum $J_B = 1$, aligned maximally parallel to μ with cylindrical symmetry (i.e. $m_B = \pm 1$ only with respect to the μ axis). This distribution, with respect to a frame with z parallel to μ, is described by $A_0^{(2)}(\mu) = +0.5$, and $A_q^{(2)}(\mu) = 0$ for $q = 1$ and 2 [for this example, we assume $a_0^{(2)}(\parallel) = a_0^{(2)}(\perp)$, so that Eq. (4.6) gives $A_0^{(2)}(\mu) = A_0^2(\text{iso}; \mu)$ and $A_0^2(\text{aniso}; \mu) = 0$]. We can calculate the description of this distribution with respect to the recoil frame by transforming by θ, the angle

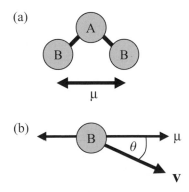

Fig. 4.2. The axial recoil photodissociation of a triatomic molecule AB_2 via a single dissociative state with μ parallel to the B–B axis, showing that there is an intermediate angle θ between $\mathbf{v}$ and μ (θ is neither $0°$ nor $90°$).

between **v** and **μ** (see Fig. 4.2b):

$$\mathbf{A}_q^{(2)}(v) = \sum_{q'=-2}^{2} d_{q'q}^2(\theta)\, \mathbf{A}_{q'}^{(2)}(\mu) \tag{4.8}$$

$$\mathbf{A}_0^{(2)}(v) = P_2(\cos\theta)\, \mathbf{A}_0^{(2)}(\mu) \tag{4.9a}$$
$$\mathbf{A}_1^{(2)}(v) = \sqrt{3/2}\,\sin\theta\cos\theta\, \mathbf{A}_0^{(2)}(\mu) \tag{4.9b}$$
$$\mathbf{A}_2^{(2)}(v) = \sqrt{3/8}\,\sin^2\theta\, \mathbf{A}_0^{(2)}(\mu) \tag{4.9c}$$

Notice that for $\theta = 0$ or $90°$ (i.e. for a pure parallel or perpendicular transition), $\mathbf{A}_1^{(2)}(v) = 0$, as expected. For $\theta = 30°$ (which is approximately the value of θ for the example described above), all the $\mathbf{A}_q^{(2)}(v)$ are nonzero. Of course, plotting the $\mathbf{A}_q^{(2)}(v)$ with respect to the recoil direction will show that the distribution is in fact cylindrically symmetric with respect to **μ** [or the $\mathbf{A}_q^{(2)}(v)$ can be transformed again as in (4.8), this time by angle $-\theta$, to recover the original values of $\mathbf{A}_q^{(2)}(\mu)$]. The point of this example is to make clear that the photofragment alignment is measured with respect to the recoil direction, **v**, and that if the alignment seems complicated in this frame, the alignment may still be expressible in a more symmetric fashion with respect to some other frame of physical significance.

4.3.3 Finite lifetime of dissociation

As mentioned in Section 2.4.1 the dissociation lifetime, τ, reduces the spatial anisotropy parameter by the factor $D(\tau)$ $[\beta(\tau) = \beta(0)\, D(\tau)]$, where $D(\tau)$ is given by [15]:

$$D(\tau) = \frac{1 + \omega^2\tau^2}{1 + 4\omega^2\tau^2}. \tag{4.10}$$

Clearly, there is no change for $\tau = 0$ $[D(0) = 1]$, whereas for long dissociation lifetimes, the spatial anisotropy is reduced to a quarter of its $\tau = 0$ value $[D(\tau \to \infty) \to 1/4]$.

Consider a prompt dissociation process that produces photofragments that are described by $\beta(0)$ and $a_q^{(k)}(p)$. Then, let us consider how the observables $A_q^k(p, \tau)$ change if the dissociation lifetime τ becomes appreciable (while also assuming that the dissociation mechanism does not change in any other way, i.e. the $a_q^{(k)}(p)$ do not depend on τ). Equation (4.5) is modified by multiplying the angular distributions by the factor $D(\tau)$, and by writing β as $\beta(0)$, because the population of the parallel and perpendicular channels is determined by $\beta(0)$ and not $\beta(\tau)$. Following through with the derivation, we see again that the photofragment laboratory distribution is described by (4.7), except that this time the $A_q^k(p)$ are replaced by the $A_q^k(p, \tau)$

5

Time resolved cameras

DANIEL ZAJFMAN, ODED HEBER AND DANIEL STRASSER

5.1 Introduction

For many years two-dimensional (2-D) and three-dimensional (3-D) fragment imaging techniques have been successfully used in the study of molecular structure [1] and for the study of the dynamics of various molecular dissociation processes, such as photodissociation [2], dissociative recombination [3], atom–molecule collision-induced dissociation [4], dissociative charge exchange [5], and others (see review by Zajfman and Heber [6]). The basic experimental scheme includes induced dissociation of a single molecule, from either a molecular ion beam or gas target, and the fully correlated measurement of the asymptotic velocity vectors of the outgoing fragments. If the initial velocity of the molecule is large, then all the fragments will be projected into a cone defined by the ratio of their transverse velocities and the initial beam velocity. In such a case, the transverse velocities are deduced from the 2-D position on the surface of a position sensitive detector, while the longitudinal velocities can be derived from the time of arrival at the detector. The specific physical information provided by the images depends on the particular dissociation process. In general, one obtains information about the initial molecular quantum state prior to the dissociation and the final state of the fragments and about the dynamics of the reaction, such as angular dependence, kinetic energy release or potential curves.

The two important features of a 3-D imaging detector are the ability to measure both time and position with good resolution, and the possibility to measure simultaneously the arrival of multiple fragments. The required resolution is typically $\sim 100\,\mu\text{m}$ in position measurement and $\sim 1\,\text{ns}$ (for keV beams or recoil experiments) or $100\,\text{ps}$ (for MeV beams) in time measurement.

The first stage of the detection and amplification of the particle impact signal is typically done by a standard micro-channel plate (MCP) detector. The MCP is usually followed by various anode schemes such as segmented anodes, cross-wire anodes [7, 8] or delay-lines [9]. With the advent of cheap and high quality charge-coupled device (CCD) cameras, direct imaging of the position of particle impact using a phosphor screen as an anode for the MCP has become more popular. These systems have the advantage of being able to measure a great number of particles simultaneously. However, for multi-particle impact, the time information is usually lost, unless a special anode is added, such as pickup wires between the MCP and the phosphor screen, or segmented-anode photomultipliers where the time can be measured via optical coupling [2, 10, 11]. In these cases, the multiplicity (i.e., the number of particles which can be detected simultaneously) is again limited by the final number of wires or anode segments. On the other hand, these systems have demonstrated very good time resolution (~ 60 ps in some cases) [10].

The essential difficulty in any readout design is the requirement to measure simultaneously correlated time and position of many fragments with good resolution. In most of the 3-D detectors, the necessary resolution can be achieved, but the number of particles which can be detected is limited. Here, we present a new scheme for a multi-particle 3-D imaging detector, which allows for the simultaneous measurement of both arrival time and position of particles, the number of particles being practically unlimited, as long as they do not hit the detector at the same position in the same event. Furthermore, this scheme has the advantage of locating the readout system outside the vacuum chamber, as only optical coupling is used. The position and time resolution achieved with such a system is presently $50\,\mu$m and ~ 1 ns respectively, but better time resolution is possible.

In the following section, we describe the basic principles of this new 3-D imaging scheme. Technical details about the first prototype we have constructed in our laboratory are given in Section 5.3, and Section 5.4 describes the analysis of the initial results related to the time calibration and resolution. Using this detector, we have carried out a preliminary experiment to demonstrate the multi-particle capability of this detector, the results of which are described in Section 5.5. In Section 5.6 we outline the possible applications of this new method.

5.2 3-D imaging scheme

The basic idea behind the new scheme for 3-D imaging is shown in Fig. 5.1. For simplicity, we describe here the working scheme using two particles, P_1 and P_2 (Fig. 5.1), but the generalization to any number of particles is trivial. Assume that the particles hit the detector at positions (x_1, y_1) and (x_2, y_2), and that their times of arrival are t_1 and t_2.

The first detection stage is made of a standard MCP detector, with a phosphor screen as an anode. Each particle impact on the MCP induces an electron shower which locally excites the phosphor screen. The excited spot on the phosphor screen

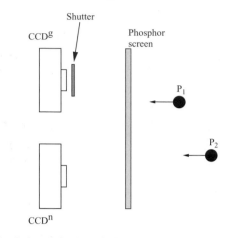

Fig. 5.1. Schematic description of the 3-D imaging method.

decays with a characteristic time (τ) which is dependent on the properties of the phosphor screen, but not on the velocity, mass or charge of the incoming particle. The second detection stage consists of two CCD cameras, CCD^n and CCD^g (Fig. 5.1) focused on the phosphor screen and a fast optical shutter in front of one of the cameras (CCD^g). This fast shutter is open until a certain time t_g, where t_g is greater than both t_2 and t_1.

The 2-D positions of the particle impacts (i.e., (x_1, y_1) and (x_2, y_2)) are obtained by digitizing the video signal of one of the cameras, usually the one without the fast optical shutter. The main innovation in the presented scheme is related to the time measurement. Let's assume for simplicity that the light emission from the phosphor decays exponentially. Any other time dependence would do as well, the only requirement being that the function is well defined.

Assuming that the CCD camera integrates over a time much longer than the decay time of the light emission, the intensity measured by the camera CCD^n for each of the particles is the integral of the light emission over the decay time:

$$I_1^n = \int_{t_1}^{\infty} i_1 e^{-(t-t_1)/\tau} dt = i_1\tau$$

$$I_2^n = \int_{t_2}^{\infty} i_2 e^{-(t-t_2)/\tau} dt = i_2\tau$$

(5.1)

where I_1^n and I_2^n are the light intensities measured by the camera CCD^n for particles P_1 and P_2 respectively, and i_1 and i_2 are the peak intensities of each light pulse produced by the phosphor screen. Figure 5.2a, b illustrate the light intensity for each spot.

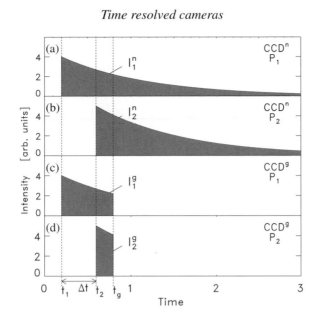

Fig. 5.2. Graphical representation of the intensities integrated by the CCD cameras: (a) CCD^n and particle P_1, (b) CCD^n and particle P_2, (c) CCD^g and particle P_1, and (d) CCD^g and particle P_2. The shaded areas represent the integrated intensities.

For the second camera (CCD^g in Fig. 5.1), which is gated by the fast shutter, the integrated light intensity for the particles is given by

$$I_1^g = \int_{t_1}^{t_g} k i_1 e^{-(t-t_1)/\tau} dt = k I_1^n \left[1 - e^{(t_1 - t_g)/\tau} \right]$$
$$I_2^g = \int_{t_2}^{t_g} k i_2 e^{-(t-t_1)/\tau} dt = k I_2^n \left[1 - e^{(t_2 - t_g)/\tau} \right], \tag{5.2}$$

where I_1^g and I_2^g are the integrated intensities measured by CCD^g for particles P_1 and P_2. The constant k is included to take into account that even though both CCDs have the same field of view, the intensity reaching them for each light spot can be different. Figures 5.2c, d explain graphically the meaning of the various quantities.

Let's define by Ω the ratio of the light intensities measured by the cameras. Then, for each particle we get the following ratios:

$$\Omega_1 \equiv \frac{I_1^g}{I_1^n} = k \left[1 - e^{(t_1 - t_g)/\tau} \right]$$
$$\Omega_2 \equiv \frac{I_2^g}{I_2^n} = k \left[1 - e^{(t_2 - t_g)/\tau} \right]. \tag{5.3}$$

The values of Ω_i are in fact the normalized intensities measured by the gated camera (CCD^g) for each particle. Inverting these functions, we can extract the times of impact t_1 and t_2:

$$t_1 = t_g + \tau \ln\left(1 - \frac{\Omega_1}{k}\right)$$

$$t_2 = t_g + \tau \ln\left(1 - \frac{\Omega_2}{k}\right)$$

$$(5.4)$$

For many practical purposes, the quantity of interest is the time difference between the impact of the particles: $\Delta t = t_1 - t_2$.

$$\Delta t = t_1 - t_2 = \tau \ln\left(\frac{1 - \frac{\Omega_1}{k}}{1 - \frac{\Omega_2}{k}}\right). \qquad (5.5)$$

This quantity depends only on the normalized intensities Ω_i, the characteristic decay time of the light intensity τ and the constant k.

The above description can be easily generalized to any number of particles hitting the detector in the same event (it is sufficient to change the indexes 1 and 2 to i and j), as long as they are spatially separated. Also, it is clear that any functional dependence for the light decay is acceptable, as long as it is well defined. However, in order to obtain resolution in the nanosecond range, a few specific points need to be clarified.

The first point is related to the dynamic range of such a detection scheme. From Fig. 5.2 it is clear that if the time separation between two particles (Δt) is much larger than the decay time τ, no practical time information can be extracted, as the intensities measured by the gated (CCD^g) and nongated (CCD^n) cameras will be practically time independent. On the other hand, there is absolutely no limit on the smallest time difference which can be measured, i.e., it is possible to measure time and position even when $\Delta t \rightarrow 0$.

A second interesting point is that if only time differences are to be measured between different particles, the gating time t_g does not need to be known (see equation 5.5). This means that any jitter in this value does not deteriorate the time resolution of the system, as long as the closing time is relatively fast ($\ll \tau$). It is the fluctuations in the time dependence of the closing function (represented, for simplicity, as a sharp edge in Fig. 5.2) which will affect the time resolution.

The last point is related to the expected time resolution. From (5.5) it is clear that an important contribution to the dispersion in the value of Δt comes from the uncertainty in the measurement of the normalized intensities Ω_i. Thus, photon statistics as well as noise in the camera and digitization systems are the most important factors. Because of the logarithmic dependence of Δt on the intensity ratios, the time resolution will not be a constant value (as it is in most time measurement schemes based on a pickup signal from an anode), but will depend on the time

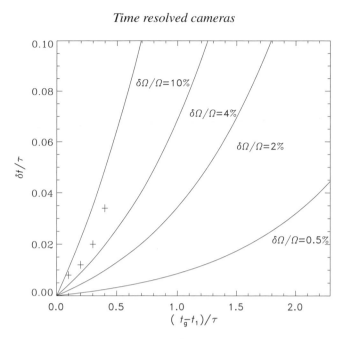

Fig. 5.3. Calculated time resolution versus the time difference between the gating time and the arrival of the particle $(t_g - t_1)$, in units of τ, for different relative widths of the distribution of intensity ratios. The crosses are the result obtained with the present setup, for $\delta\Omega/\Omega = 4\%$).

difference itself $(t_g - t_i, i = 1, \ldots, n$ where t_i is the time of arrival of particle i). Figure 5.3 shows the expected time resolution as a function of the time difference between the gating time t_g and the time of arrival of the particle $(t_g - t_1)$, in units of τ, for different widths in the distribution of the intensity ratio measurement $(\delta\Omega/\Omega)$. For example, using a phosphor screen with a decay time $\tau = 50$ ns and an imaging system (CCDs and gated image intensifier) which can measure the intensity ratio with a precision of $\delta\Omega/\Omega = 4\%$, we obtain a time resolution of $\delta t/\tau = 0.07$, which is equivalent to $\delta t = 3.5$ ns for a particle arriving 50 ns before the optical gate closes $(t/\tau = 1$ on the horizontal axis). Using the same phosphor, but a time difference $(t_g - t_1) = 10$ ns $(t/\tau = 0.2)$, changes the time resolution to $\delta t/\tau = 0.0075$ (i.e., 375 ps at 10 ns). Hence, the best resolution is obtained when the optical gate is closed as soon as possible after the particles have reached the detector. However, it must be pointed out that the above description assumes that the error in the ratio of intensities Ω is independent of t_g, and that the contributions from other factors can be neglected (see Section 5.4). The curves shown in Fig. 5.3 should be taken as a basis for evaluating the functional dependence of the resolution and to give an idea of the effective dynamic range of the system. In the following, we describe in detail a working prototype of a detector based on the concept described above.

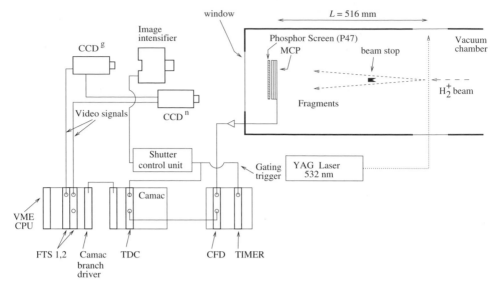

Fig. 5.4. Schematic view of the experimental setup for the detector.

5.3 Experimental setup

Figure 5.4 shows the experimental setup of the new detector. The detection stage, located inside a vacuum chamber, consists of a Z-stack of MCPs, 40 mm in diameter, coupled to a P47 (Y_2SiO_5:Ce) phosphor screen. The P47 phosphor screen has a characteristic decay time of 50 ns. The light emission is in the blue and peaks at $\lambda = 400$ nm. Two standard, black and white CCD cameras (model CS8310B, Teli) are located outside the vacuum chamber. The optical shutter is a gated image intensifier (model ESCU-10D, Photek) with a 2 ns closing time. This image intensifier is a stand-alone device, which is made of a photocathode connected to a fast switch, an MCP and a P20 phosphor screen. Standard C-mount lenses are adapted to the CCDs and to the photocathode side of the image intensifier. Both the camera CCD^n and the image intensifier are focused directly on the phosphor screen of the MCP detector, while camera CCD^g is focused on the phosphor screen of the image intensifier (25 mm diam.). The two cameras are synchronized so that the output video signals contain information about the same event. The video signals are sampled and digitized by two homemade frame grabbers, located in a VME computer, which are used as frame threshold suppressors (FTS) [10]. Each FTS unit samples the video signal with a fast 8-bit ADC, and stores, in an on-board memory, the pixels' locations (x_j, y_j) and their intensities i_j, but only if i_j is above a preset threshold. Images are sampled at a rate of 25 frames/s, with a resolution of 600×256 pixels, but the average number of pixels stored in memory for each light spot on the phosphor screen is ~ 100. An online real-time analysis program uses a peak finding procedure to determine the centre of mass average positions and integrated intensities (I_i^n and I_i^g, see (5.1) and (5.2), for $i = 1, \ldots, n$) of each particle image in both cameras.

For calibration purposes, for which single particles are used, a fast amplifier picks up the pulse induced on the back of the MCP. The pulse is fed into a constant fraction discriminator (CFD) which provides a start signal for a time-to-digital converter module (TDC). The TDC is stopped by the trigger signal to the image intensifier, providing a measure of the closing time t_g. The trigger signal itself is provided by an external event, such as a fast chopper pulsing the beam, or the firing of a laser which dissociates an incoming molecular ion. Using directly the signal coming from the MCP as a trigger for the shutter was found to be impossible in our system because of the large delay ($\sim 60\,\text{ns}$) existing in the shutter control unit between the trigger input and the actual time the photocathode is switched off.

5.4 Detector time calibration

As pointed out in Section 5.2, the major uncertainty to the time measurement is related to the intensity ratio between the two cameras, for each light pulse on the phosphor screen. Using a 4.2 keV Ar^+ beam produced from an electron impact ionization source, we have measured this ratio for various settings of the MCP, phosphor and image intensifier voltages. For this measurement, the optical shutter is kept open all the time. Figure 5.5 shows the ratio of intensities between CCD^g and CCD^n (I_1^g/I_1^n for $t_g \to \infty$), which is equal to the constants k (see 5.3). The relatively narrow peak with a width $\sigma = 4\%$ demonstrates that (1) the imaging system, including the CCDs and the FTSs is stable and not very noisy, and (2) the image intensifier conversion between the incoming number of photons on its photocathode and the light output from its phosphor screen is linear and has a narrow distribution. In fact, it was found that the two CCDs and frame grabbers are responsible for 75% of the distribution width shown in Fig. 5.5, while the image intensifier produces an additional 25%. This is valid over a wide range of intensities produced by the P47 phosphor screen, except for the very low intensity pulses, which represent a few percent of the total number of events.

The time calibration for the detector is in fact a measurement of the light emission of the phosphor screen as a function of time, which in Section 5.2 has been described by the characteristic decay time τ. To measure this function, we have used single particle impact (4.2 keV Ar^+) on the detector, and calibrate the ratio of integrated intensities I_1^g/I_1^n as a function of t_g. The value of t_g was given by the TDC (see Fig. 5.4) which measured the difference between the time of impact and the trigger to the shutter control unit of the image intensifier.

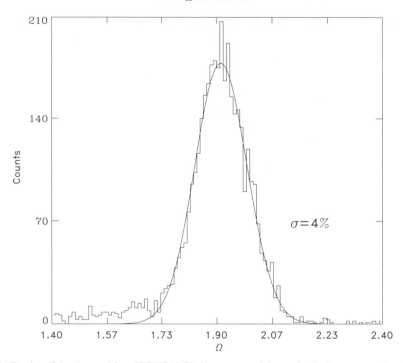

Fig. 5.5. Ratio of the intensities (CCD^g/CCD^n) measured for a single impact on the detector. The measurement is made with the gated image intensifier always open. The curve is a Gaussian fit to the measured distribution, with a relative width $\sigma = 4\%$.

Figure 5.6 shows a scatter plot where the measured ratios of intensities $\Omega = I_1^g/I_1^n$ is plotted versus $(t_g - t_1)$. The functional dependence observed is consistent with (5.3), which is shown as a solid line using the decay time $\tau = 50$ ns. This result is in good agreement with the value quoted in the literature for a P47 phosphor [12]. The functional dependence was found to be independent of the position of impact on the screen, the anode acceleration voltage and the energy and type of particles. Identical results were obtained with a H_2^+ beam at 4.2 keV, as well as a proton beam at 2.1 keV. As can be seen from Fig. 5.6, at short times ($(t_g - t_1) < 30$ ns), a linear approximation is valid to describe the relation between the ratio of intensities and the time. For time shorter than 2 ns, a slight departure from the expected dependence is observed, which is probably due to the fast rising time of the phosphor light. The dispersion of points around the curve gives an indication of the time resolution one can expect with the present system. Also, it is evident that the time resolution gets worse when the time difference between the impact of the particle on the detector and the closing time of the shutter ($t_g - t_1$) gets large. This is in complete agreement with the model described in Section 5.2 (see Fig. 5.3).

For short times ($(t_g - t_1) < 10$ ns) the time resolution seems to be independent of t_g, which may indicate that other parameters are limiting the time resolution.

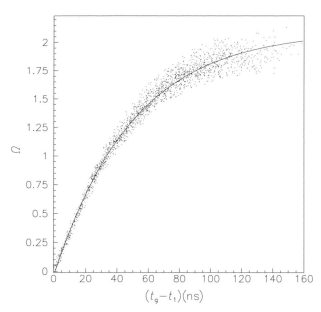

$(t_g - t_1)(\text{ns})$

Fig. 5.6. Measured ratio of integrated intensity versus the difference between the closing time of the gate t_g and the time of arrival t_1 of particle P_1.

Among them is the final resolution to which the real gating time could be measured: Although the knowledge of t_g is not needed for measuring time difference between the impact of various particles (see (5.5)), a good knowledge of it is needed for the calibration. If jitter exists between the trigger of the shutter control unit and the actual closing time of the image intensifier, it will influence the ultimate measured resolution of the system, and could explain why the measured resolution is independent of time for very short $(t_g - t_1)$. This does not imply that the true resolution of the detector when measuring the time difference between the impact of several particles is not improving with decreasing $(t_g - t_1)$, but the present system does not allow precise measurement of this quantity.

In order to quantify the measured time resolution, we have sliced the scatter plot shown in Fig. 5.6 using narrow horizontal bins, and projected the data on the time axis around different $t_g - t_1$ values. For each of these slices, a Gaussian distribution was obtained, whose width is the time resolution around that specific time difference. Figure 5.7 shows the distribution obtained for four different time slices: (a) $0 < (t_g - t_1) < 5$ ns, (b) $5 < (t_g - t_1) < 10$ ns, (c) $10 < (t_g - t_1) < 15$ ns, and (d) $15 < (t_g - t_1) < 20$ ns. These spectra were fitted using Gaussian distributions, their widths being representative of the typical resolutions obtained in these time intervals. Figure 5.3 shows a comparison between the expected resolution, taking into account only the width of the intensity ratio, and the values obtained with the present detector ($\delta\Omega/\Omega = 4\%$). It can be seen that the functional dependence is

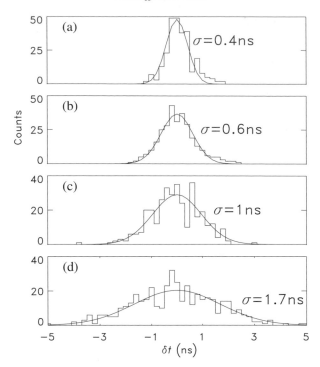

Fig. 5.7. Relative time resolution for different regions of $(t_g - t_1)$ around an average value $<t>$ which is in the middle of the following ranges: (a) $0 < t_g - t_1 < 5$ ns, (b) $5 < t_g - t_1 < 10$ ns, (c) $10 < t_g - t_1 < 15$ ns, and (d) $15 < t_g - t_1 < 20$ ns. The solid line is the result of a Gaussian fit.

similar to the predicted one, although the final resolution is worse than calculated. The difference can be due to various factors (see Section 5.2). Among them, the exact closing time which is needed for the resolution measurement, although this is not necessary if only the time difference is required. Also, the value of $\delta\Omega/\Omega$ may be different (worse) for short shutting times, as less photons are integrated by CCD[g]. However, the general behaviour of the system is well understood.

The above values were obtained with the detector shown in Fig. 5.4, which consists of simple CCD cameras and simple optics. A better and more rugged system should improve the time resolution. Also, direct coupling of the shutter to the camera, and not through an additional lens, could enhance the time resolution of the system.

5.5 Multi-particle detection

In the previous section, we have demonstrated that the new scheme for 3-D imaging can provide correlated position and time information for a single particle with good position and time resolution. In the present section we show the preliminary results of a simple test we have carried out to measure the position and time of two correlated particles hitting the detector. We have chosen to measure the kinetic energy release in the photodissociation of H_2^+ ($H_2^+ + h\nu \rightarrow H + H^+$). Such an experiment was already carried out by Helm and Cosby more than 10 years ago [13] in order to measure the rovibrational energy and populations of H_2^+ in a beam. Using a position and time sensitive detector based on a pair of MCP detectors and a specially designed anode, they have observed the spectrum of kinetic energy release due to the photodissociation process.

For H_2^+ produced in an electron impact ionization source, the number of populated vibrational states is rather large. For vibrational states $v > 6$, photodissociation with a photon energy of $h\nu \sim 2\,\text{eV}$ is possible. The transition occurs between the ground electronic potential ($X^2\Sigma_g^+$) and the dissociative state $A^2\Sigma_u^+$ [13]. For a molecular ion initially in a vibrational state v and a rotational state N, the kinetic energy release E_k is given by the relation

$$E_k = h\nu - D_{v,N} \tag{5.6}$$

with $D_{v,N}$ being the dissociation energy of the bound rovibrational level with respect to the dissociation limit to which the continuum energy is referenced. The cross-section for photodissociation of a bound level (v, N) is, among other parameters, a direct function of the overlap between the initial rovibrational wave function and the continuum wave function. Thus, the kinetic energy release spectrum should be made of a series of peaks for each (v, N) with intensities depending on the initial population and the wave function overlap. The rotational state distribution is characterized by the temperature of the parent gas in the ion source H_2 ($\sim 1500\,\text{K}$). More details can be found in Ref. [13]

In the present case, a pulsed beam of photons from a Nd:YAG laser operating at $\lambda = 532\,\text{nm}$ ($h\nu = 2.33\,\text{eV}$) was used in order to photodissociate the 4.2 keV beam of H_2^+. The interaction point was located at a distance of 516 mm from the surface of the MCP detector (see Fig. 5.4), and both the ion beam and the laser intensities were chosen so that less than one event was produced per pulse. A beam stop was used to block the H_2^+ beam at a distance of 266 mm from the detector. The polarization of the laser was chosen to be perpendicular to the beam direction. Since the photodissociation process is characterized by a parallel transition moment, this choice of orientation in the laser polarization results in an angular distribution of the photofragments which is given by $\cos^2 \theta$, when θ is the angle between the internuclear axis and the polarization vector.

The trigger to the gated image intensifier was produced by a delayed signal from the laser pulse, which was set so that the closing time t_g would be about 20 ns after the centre-of-mass of the fragments would hit the detector. The position (x_i, y_i) was determined from the data of CCD^n. For each of the two particles, the intensity ratios were extracted from the correlated CCD frames, and the time of arrival of each particle was obtained from the calibration curve (Fig. 5.6). Since the maximum time difference between two fragments is ~ 30 ns, it means that the shutter was closed at a maximum time delay of ~ 35 ns relative to the first particle. As pointed out in Section 5.2, it is the time resolution of the earliest particle which forms the dominant contribution to the overall time resolution. Based on these values, a time resolution of ~ 2 ns can be expected (see Fig. 5.3). Because of the small difference in energy between the vibrational states of H_2^+ involved in this reaction (~ 100 meV), such a time resolution would wash out the structure in the kinetic energy release. Thus in order to improve the resolution in the kinetic energy release spectrum, we have kept only events with small Δt, so that the contribution of the timing resolution to the kinetic energy release was minimized. Figure 5.8 shows the kinetic energy release spectrum obtained only for events arriving within a time difference $\Delta t < 2$ ns (i.e., those molecular ions for which the internuclear axis was parallel to the detector plane upon dissociation).

The arrows shown in Fig. 5.8 indicate the expected position for the kinetic energy release for the various vibrational states v, for rotationally cold ($N = 0$) molecular ions. The positions of the measured peaks are in good agreement with the expected kinetic energy release. The width of the peaks is mainly due to the rotational temperature. A detailed analysis of this spectrum can be found in the work of Helm

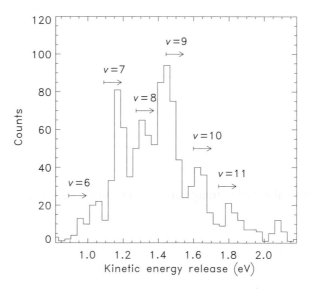

Fig. 5.8. Kinetic energy release for the photodissociation of H_2^+ by 532 nm photons. The arrow indicates the expected range of kinetic energy release for the various vibrational states v, assuming a rotational temperature of $T = 1500$ K.

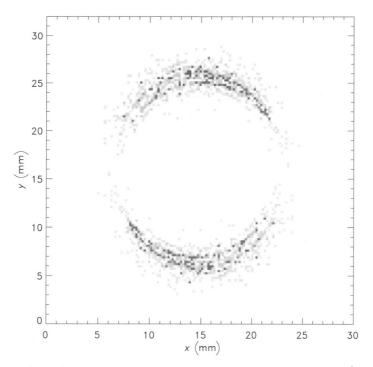

Fig. 5.9. Two-dimensional image as measured for the photodissociation of H_2^+. Only events for which $\Delta t < 2\,\text{ns}$ are shown.

and Cosby [13]. Figure 5.9 shows the 2-D data as measured by CCD^n. The effect of the vertical polarization is clearly seen on the photodissociation pattern. Here again, only events for which $\Delta t < 2\,\text{ns}$ are shown.

5.6 Summary

The new 3-D detection scheme we have developed allows for the correlated measurement of time and position of a basically unlimited number of particles. The spatial resolution achieved with the prototype we have designed is $50\,\mu\text{m}$ while the time resolution depends on the difference in time between the particle impact and the closing of the gated image intensifier, and varies between 0.4 and 2 ns for a time difference of 2 and 30 ns, respectively. We have demonstrated that the time resolution is limited mainly by the width of the intensity ratio as measured by the imaging system. We have reached a limit of $\delta\Omega/\Omega = 0.04$ for the ratio of integrated intensities with the present simple hardware, but improvement is readily achievable. The quality of the CCD cameras as well as of the frame grabbers can be improved so that the width of the intensity ratio measurement would be of the order of 0.5%. This would lead to a time resolution well below the 1 ns for a phosphor screen with a decay time of 50 ns, over the whole 0–50 ns range. The dynamic range of the time measurement is a function of the characteristic decay time of the phosphor screen, which can be adapted for a specific application. Decay times in the range of a few nanoseconds (as with the P47 type used here) up to a few milliseconds (for the P11 phosphor) are available, thus spanning a wide range of applications. We have demonstrated one application using a simple molecular dissociation process, but we believe that the present detector opens up a wide range of new experiments which could be carried out without further conceptual development. Three-dimensional imaging of the dissociation process of large molecules and clusters is now possible, and multiparticle correlation in recoil experiments can be made without any limit to the number of recoil fragments. We hope that this technique will open new possibilities in other fields, and among these, we point out the tomographic atom probe [14] where 3-D images of single atomic layers can be obtained via field evaporation of surface atoms, and time-of-flight measurement of evaporated ions. In this case, many ions impact on the detector at the same time, and their correlated positions and times need to be measured simultaneously. A rather important implication of the present method is the possibility to use it for 3-D vision (with photons hitting the detector, instead of particles). In such a case, a full 3-D image can be recorded with a single pulse of light. A patent application has been filed for this technique.

Acknowledgements

This work was supported by the German Ministry of Education, Science, Research and Technology (BMBF) within the framework of the German-Israeli Project Cooperation in Future-Oriented Topics (DIP), and by the Heineman Foundation. A grant from the Levine Foundation is also acknowledged.

References

1. Z. Vager, R. Naaman, E. P. Kanter, *Science* **244**, 426 (1989).
2. A. J. R. Heck, D. W. Chandler, *Annu. Rev. Phys. Chem.* **46**, 335 (1995).
3. D. Zajfman *et al.*, *Phys. Rev. Lett.* **75**, 814 (1995).
4. V. Horvat, O. Heber, R. L. Watson, R. Parameswaran, J. M. Blackadar, *Nucl. Instrum. Methods* **B99**, 94 (1995).
5. W. J. van der Zande, W. Koot, D. P. de Bruijn, *Phys. Rev. Lett.* **57**, 1219 (1986).
6. D. Zajfman and O. Heber. In *Accelerator-Based Atomic Physics Techniques and Applications*, eds. S. M. Shafroth, J. C. Austin, (American Institute of Physics, Woodbury, New York, 1997), pp. 481–507.
7. C. Martin, P. Jelinsky, M. Lampton, R. F. Malina, H. O. Anger, *Rev. Sci. Instr.* **52**, 1067 (1981).
8. D. P. de Bruijn, J. Los, *Rev. Sci. Instr.* **53**, 1020 (1982).
9. R. Dörner *et al.*, *Phys. Rev. Lett.* **76**, 2654 (1996).

10. D. Kella *et al.*, *Nucl. Instrum. Methods* **A329**, 440 (1993).
11. Z. Amitay, D. Zajfman, *Rev. Sci. Instrum.* **68**, 1387 (1997).
12. Hamamatsu Photonics, *Image intensifiers products* catalog, 1991.
13. H. Helm, P. C. Cosby, *J. Chem. Phys.* **86**, 6813 (1987).
14. D. Blavette, B. Deconihout, A. Bostel, J. M. Sarrau, M. Bouet, A. Menand, *Rev. Sci. Instrum.* **64**, 2911 (1993).

6

3-D Imaging technique – observation of the three-dimensional product momentum distribution

ALEXEI I. CHICHININ, TINA S. EINFELD,
KARL-HEINZ GERICKE AND CHRISTOF MAUL

6.1 Introduction

As we have seen in the previous chapters inversion algorithms are required in conjunction with two-dimensional (2-D) detection methods in order to reconstruct the velocity (speed and angle) distributions of the products of photodissociation processes. Since these generally need to make certain assumptions about the symmetry of the measured distributions it would be desirable to measure the velocity distribution directly by making a simultaneous measurement of the position and arrival time of each of the photoproducts. In the previous chapter we saw how two charge-coupled device (CCD) cameras can be used in conjunction to make such a measurement. The present chapter develops this idea further by describing a newly developed three-dimensional (3-D) photofragment imaging technique. In this context the term '3-D imaging' refers to the simultaneous measurement of all three coordinates of a single particle, which are defined by the spatial position in the 2-D surface of the position-sensitive detector (PSD) and by the time of arrival at the detector (the third dimension) of the ionized product of a photodissociation process. The transverse velocity components (v_x, v_y) of the initial velocity of the product are determined from the measured 2-D impact position on the PSD surface, while the measured time of arrival gives the longitudinal component (v_z) of the velocity. Hereafter the laboratory axes X, Y, and Z are directed along the laser beam, the molecular beam, and the accelerating electric field, respectively (Fig. 6.1). Directly determining the 3-D velocity distribution of the product provides complete information about the photodissociation process. Note that this technique may easily be extended to study the velocity distributions of the products of chemical reactions; a second molecular beam must be added in this case.

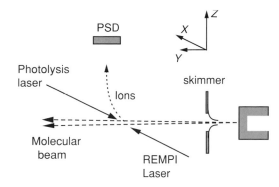

Fig. 6.1. Schematic depiction of the REMPI/TOF mass spectrometer with PSD provided for 3-D photofragment imaging. Also shown is the coordinate system used in this work: the X axis is defined by the laser beam, the Y axis by the molecular beam, and the Z axis by the spectrometer axis.

6.1.1 Why 3-D imaging?

What are the principal advantages of the 3-D imaging techniques over the 2-D 'velocity map imaging' method? As we have seen in Chapter 3 in order to analyse 2-D 'velocity map' images the Abel inversion algorithm or the forward convolution analysis is usually used. The Abel algorithm is rather simple, but it has several limitations: first, it assumes cylindrical symmetry of the initial 3-D distribution, second, the electric field vector $\mathbf{E}$ of the dissociative radiation must be perpendicular to the Z axis ($\mathbf{E} \perp \mathbf{Z}$), and third, a rather large number of data points is necessary to reconstruct the 3-D distribution of the ions. The forward convolution analysis is more powerful, but it is computationally expensive and, in general, does not lead to unique solutions [1].

In contrast to the 2-D method, the 3-D imaging approach is quite straightforward. Integration techniques, such as the Abel inversion algorithm or the forward convolution analysis, are not required. Moreover, the 3-D imaging approach is more universal: it can be used with any polarization geometry, including the common cases ($\mathbf{E} \parallel \mathbf{Z}$) and ($\mathbf{E} \perp \mathbf{Z}$), and it does not assume any symmetry of the initial 3-D distribution.

There are several phenomena that may break the cylindrical symmetry of the initial fragment distributions. First, alignment of the parent molecules in the cold molecular beam: collisions occurring during the supersonic expansion can produce the molecules in their lowest rovibrational states with alignment of the rotational angular momentum along the axis of the molecular beam [2,3]. The degree of alignment, which can be rather large, varies strongly with the molecular speed, and it is expected to depend on the experimental conditions [2]. The phenomenon has been observed for O_2 [4,5] and N_2 [6] seeded in lighter carriers, for CO seeded in

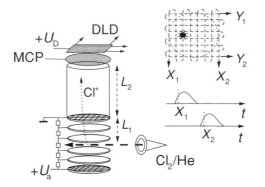

Fig. 6.3. Left: a schematic diagram of the TOF MS. A homogeneous electrostatic field in the acceleration region (L_1) is created by a system of electrode rings. The system is confined by metallic meshes from both sides. Behind the drift region (L_2) the ions impinge upon the MCP. Secondary electrons from the MCP are accelerated towards the DLD. Right: illustration of the DLD principle. Every event generates four signals on the DLD: one on every end of the two orthogonal delay lines, denoted X_1, X_2, Y_1, and Y_2. X and Y coordinates are obtained from the differences $X_2 - X_1$ and $Y_2 - Y_1$, respectively.

is given by $L_S = 0.67\, d\, (P_0/P_b)^{1/2}$, where d is the diameter of the nozzle, P_0 is the initial pressure inside the nozzle and P_b is the background pressure in the jet chamber (typically 10^{-7}–10^{-10} bar). L_S ranges typically from a few centimetres to a few metres, depending on the experimental conditions.

6.2.3 TOF mass spectrometer

In the ionization chamber the skimmed molecular beam is intersected by the probe and the photolysis laser beams whose propagation direction (X axis) is perpendicular to the molecular beam propagation axis (Y axis). The ionization chamber is shown in Fig. 6.3. Products generated in the photodissociation are state-selectively ionized by the probe laser and are accelerated along the Z axis towards the field-free drift region of the TOF MS. Typically, electric field strengths of 3000 V/m are realized by applying an acceleration voltage of $U_a \approx 300$ V to the acceleration stage. The space-focusing condition $2L_1 = L_2$ should be obeyed in order to simplify the interpretation of the 3-D PSD images. Here, L_1 is the length of the acceleration region and L_2 is the length of the drift region (Fig. 6.3). Single ions are detected by a micro-channel plate (MCP) assembly (typical diameters of 4–8 cm). Subsequently, the resulting secondary electrons are monitored by the PSD.

6.2.4 Delay-line position-sensitive detector

In this section we describe the special case of a PSD, which we call a delay-line detector (DLD). The DLD consists of a delay-line anode introduced into the ionization chamber right behind the MCP assembly (Fig. 6.3). An additional potential ($U_D \approx 300$ V) applied between the MCP and the DLD accelerates the electrons to the delay-line anode. Note that the MCP signal itself, produced by incoming ions and picked up from the charging circuit of the MCP assembly, is only used for adjustment purposes and for the measurement of the mean TOF averaged over all observed single ion events, whereas all quantitative information on the 3-D velocity components of each individual ion is obtained from the DLD signal.

The idea behind a DLD is the following: a signal induced somewhere on a delay line (basically nothing more than a long wire) propagates in both directions towards the ends of the line where impedance adjusted circuits pick it up for further processing (Fig. 6.3). By measuring the time period between the signal arrival times at both ends of the delay line one can determine the position of the signal source along the delay line.

The technical realization of the DLD employed here is described in detail in Refs [18–20]. Briefly, it consists of two individual delay lines oriented orthogonal to each other, thus forming the X–Y plane (Fig. 6.4). A metal body supports ceramic rods placed on the edges; and the delay lines are wound helically on this 8×8 cm^2 supporting plate. By this folding technique, a propagation delay of 20 ns/cm and a total single-pass delay of 150 ns is realized, corresponding to a physical length of the delay line of 45 m. Each delay line consists of a pair of wires wound parallel to each other, with a small potential difference (~ 30 V) applied between the two wires of each pair. Thus, the incoming charge cloud from the MCP induces a differential signal on each delay-line pair that propagates to the delay line ends where it is picked up by a differential amplifier.

6.2.5 DLD data analysis

The output charge from the MCP resulting from each single incoming ion or photon ('event') produces altogether four differential signals, two on each delay-line pair

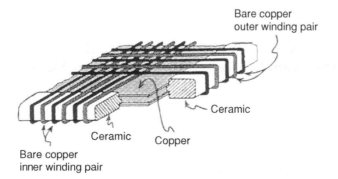

Fig. 6.4. Detail of a helical wire delay-line anode.

(Fig. 6.3). These signals are decoupled from the DC voltages on the wires, amplified, and transmitted to constant fraction discriminators and then to time-to-digital converters. Finally, one event produces two pairs of times, t_{X_1}, t_{X_2} and t_{Y_1}, t_{Y_2} on the delay lines that are wound along the X axis and the Y axis, respectively. The X and Y coordinates of a single event in time units may be calculated as $(t_{X_1} - t_{X_2}, t_{Y_1} - t_{Y_2})$ and the time of the event (corresponding to the Z coordinate) may be calculated as either $(t_{X_1} + t_{X_2})/2$ or $(t_{Y_1} + t_{Y_2})/2$. Thus, the DLD yields the 3-D coordinates of each single event.

The latter condition makes it possible to distinguish between true and false events: the event time provided by the different delay lines must coincide, hence $t_{X_1} + t_{X_2} = t_{Y_1} + t_{Y_2}$. Only those events that obey this condition are counted, and all others are ignored. If more than one ion per laser pulse strikes the MCP assembly, then each delay line produces a series of pulses and the criterion $t_{X_1} + t_{X_2} = t_{Y_1} + t_{Y_2}$ allows one to correctly assign individual pulses to individual ions.

6.2.6 Time and space resolution of the spectrometer

The space resolution of the REMPI/TOF/DLD spectrometer may be determined by experiments with a fine metallic mesh placed just in front of the MCP assembly, which is irradiated by α particles. The wires of the mesh appear as a geometric projection on the DLD, and the dimensions of the wires and the holes of the mesh allow one to determine the spatial resolution of the detector. The image of such a mesh on our DLD is shown in Fig. 6.5. From the analysis of the image one obtains a space resolution of the DLD, which can be better than 100 µm [19]. In the present experiment, the resolution was 400 µm, which still corresponds to a spatial resolution of 1 part in 200 because of the detector size of 80 mm diameter.

The DLD provides the possibility of detecting several products per laser shot, provided a minimal difference between arrival times of 17 ns is maintained. This detector dead time is mainly determined by the typical duration of the pulses on the delay lines. The time resolution for the time of an individual event in the REMPI/TOF/DLD spectrometer is usually limited by the duration of the probe laser pulse, which is of the order of 3–5 ns if a Nd:YAG pump laser is used.

6.2.7 Why DLD?

As has been described in previous chapters the common PSD for a velocity mapping experiment uses a phosphor screen as an optically collecting anode combined with read-out by video camera and/or charge-coupled device (CCD) camera [1,21]. This 'classical' method is optimized for the 2-D imaging of single particles.

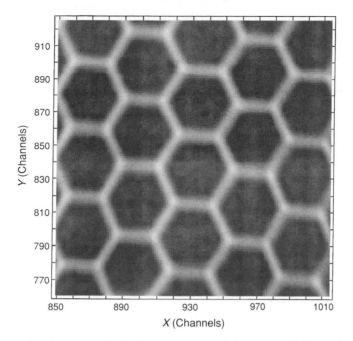

Fig. 6.5. Detail of an image acquired with the DLD [19]. The lines of the mesh were 0.1 mm in diameter. X and Y are given in channel numbers (40 channels = 4 mm).

However, it is not very useful for obtaining the remaining space coordinate via timing information due to the relatively slow electronic read-out of the optical image, and because after each laser shot a complete optical image, which must contain just one event, has to be analysed. A more promising way of using a phosphorescent detector with two cameras was described in the previous chapter by Zajfman, Hebert and Strasser, where fluorescence lifetime measurements of the phosphor screen were used to extract information of the arrival time of each particle. However, in this time-resolved camera the phosphor needs to be selected in accordance with the arrival time distribution of the particles of the process under study. The construction of a DLD is simpler and less expensive than the special imaging CCD cameras and it can be used without modification for different chemical processes.

6.3 Applications of the delay-line detector

6.3.1 Speed and temperatures of the molecular beam

Although the real advantage of the delay-line detector is its ability to determine 3-D product momentum distributions, our technique also provides an excellent

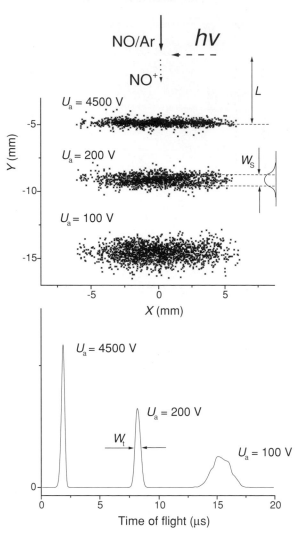

Fig. 6.6. Upper panel: space distribution of NO^+ ions from the 226.2 nm ionization of NO at different acceleration fields. The molecular beam propagates vertically from top to bottom and is intersected by the laser beam at a right angle. The position L of the molecular beam image on the DLD is the product of the average speed of the molecular beam with the time-of-flight. W_s is a measure of the longitudinal speed distribution of the molecular beam. Lower panel: distributions of the times-of-flight for the same three groups of ions. For illustrative purposes the widths W_t of the lines are increased by a factor of 100.

opportunity to study molecular beam properties. Figure 6.6 shows examples of the space distribution of NO^+ ions at different acceleration voltages that are equivalent to different times-of-flight. In these experiments, a diluted (10^{-4}) mixture of NO in $X(X = Ar, He, N_2)$ was expanded through the nozzle and NO was detected by

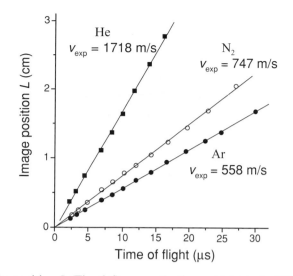

Fig. 6.7. Plot of the position L (Fig. 6.6) versus the time-of-flight t for NO/He, NO/N$_2$, and NO/Ar yields straight lines the slopes of which are the speeds v_{exp} of the molecular beams.

(1 + 1) REMPI at 226.2 nm. The analysis of the space distributions of NO along the Y axis has shown that the distribution has a Gaussian profile:

$$W(y) = W_0 \exp\left(-\frac{4\ln 2(y - L)^2}{W_s^2}\right) \tag{6.1}$$

where y is the space coordinate along the molecular beam axis, L is the position of the image of the molecular beam with respect to the laser beam–molecular beam interaction region, and W_s is the full width at half maximum (FWHM) of the space distribution. The variations of L and W_s with time-of-flight t are shown in Figs. 6.7 and 6.8, respectively, and can be used to calculate the speeds v_{exp} of the molecular beam and the translational temperature T_s.

The method to obtain the speeds v_{exp} of Ar, He, and N$_2$ molecular beams is quite straightforward: the speeds are the slopes of the linear fits to the data in Fig. 6.7, $v_{exp} = dL/dt$. The maximum theoretical speed of the molecular beam v_{max} may be calculated as

$$v_{max} = \sqrt{\frac{\gamma}{\gamma - 1}\frac{2kT_0}{m}} \tag{6.2}$$

where m is the mass of the buffer gas molecule (He, Ar, N$_2$, or H$_2$ in our case), k is the Boltzmann constant, $T_0 = 295\,$K is our room temperature, and γ is the ratio of heat capacities, C_P/C_V. The γ parameter is 5/3 and 7/5 for an atomic and

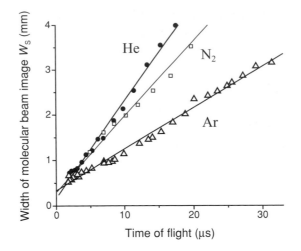

Fig. 6.8. Plot of the longitudinal width W_s of the image of the molecular beam (Fig. 6.6) versus the time-of-flight t for NO/He, NO/N$_2$, and NO/Ar. The data exhibit the expected linear dependence the slopes of which give the temperatures of the molecular beams, see equation (6.5).

a diatomic buffer gas, respectively. Equation (6.2) assumes that all rotational and translational energies of the molecule are transformed to directed, collective kinetic energy of the beam: the colder the beam, the closer v_{max} and v_{exp} values. Note that the internal temperature of the beam T_{calc} may be obtained from the difference between these speeds:

$$T_{calc} = T_0 \left[1 - \left(\frac{v_{exp}}{v_{max}} \right)^2 \right]. \tag{6.3}$$

The experimentally observed v_{exp} and the calculated v_{max} values are listed in Table 6.1.

Figure 6.8 is an experimental plot of the width W_s as a function of the time-of-flight t. The Gaussian profile of the space distribution is due to the one-dimensional Gaussian speed distribution over the Y axis, which is characterized by the temperature T_s:

$$W(v_y) = W_y \exp \left(-\frac{M(v_y - v_{exp})^2}{2kT_s} \right) \tag{6.4}$$

where M is the mass of the NO molecule. Substituting $v_y - v_{exp} = (y - L)/t$ one can find the relation between W_s and T_s by comparing equation (6.4) with (6.1):

$$T_s = \frac{M \left(\frac{dW_s}{dt} \right)^2}{8k(\ln 2)^2} \tag{6.5}$$

Table 6.1. *Summary of temperatures and speeds of the molecular beams for different buffer gases. v_{max} is the maximum beam velocity which has been calculated assuming complete cooling to a temperature of 0 K, calculated with (6.2); v_{exp} is the experimentally observed value; T_s is the translational temperature of the molecular beam derived from the spread in the v_y velocity component (in beam direction), extracted from the broadening of the impact position on the DLD, see (6.5); T_{rot} is the rotational temperature of the NO molecule extracted from spectroscopic measurements of the intensities of rotational transitions; T_{calc} is the internal temperature of the beam obtained from the difference between v_{exp} and v_{max}, (6.3). The negative calculated temperature for the Ar beam is a consequence of taking the difference of two large quantities with small experimental uncertainties*

Gas	v_{max} (m/s)	v_{exp} (m/s)	T_s (K)	T_{rot} (K)	T_{calc} (K)
Ar	554	558 ± 4	3.5 ± 0.5	3.0 ± 0.5	-4 ± 5
He	1751	1718 ± 12	12.3 ± 2	8 ± 2	11 ± 5
N_2	783	747 ± 12	12.7 ± 2	12 ± 2	26 ± 9
H_2	2931	2041 ± 40			152 ± 6

where dW_s/dt is the slope of the linear fit to the data in Fig. 6.8. The experimental results on T_s are summarized in Table 6.1. Note that the variations of W_t with TOF t can be used to calculate the translational temperature of the beam along the Z axis, where W_t is the width of the TOF distribution (Fig. 6.6). This temperature was found to be very small (about 0.2 K for NO/Ar). Thus, using this simple analysing technique, we can gain information on the beam temperature parallel and perpendicular to the propagation direction of the beam.

6.3.2 Spectroscopy of cold molecules

Another interesting application of the DLD is the REMPI spectroscopy of cold molecules. It is well known that in ultra-cold supersonic beams a considerable simplification of molecular spectra can be achieved. Here, one problem is the contribution of room-temperature background molecules and impurities which make the spectra more difficult to analyse. In time-of-flight experiments with a perpendicular geometry of molecular beam and spectrometer axis it is easy to distinguish between contributions from the molecular beam and background contributions since the particles which belong to the molecular beam are located on a very small area of the DLD which is shifted from the axis of the laser by the distance $L = v_{exp}t$. Here t is the TOF and v_{exp} is the experimentally determined speed of the molecular beam (Fig. 6.9).

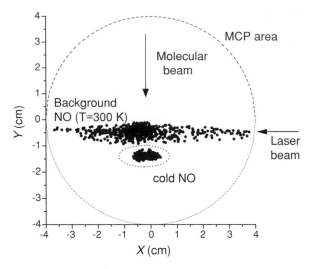

Fig. 6.9. Space distribution of NO^+ ions from the 226.2 nm ionization of NO. Ions from the molecular beam are located in a small area marked by the dotted ellipse. Only these ions are used to construct the rotational REMPI spectrum of NO.

This method allows, for example, the determination of the rotational temperature T_{rot} of NO in Ar, He, and N_2 molecular beams (Fig. 6.10). In general, there exists a good agreement between the temperatures T_{rot}, T_{exp}, and T_s as summarized in Table 6.1.

6.4 Classic photofragment imaging: photolysis of Cl_2

6.4.1 Experimental results

The quality of the 3-D imaging technique was examined by studying the well-known photolysis of molecular chlorine, $Cl_2 + h\nu$ (355 nm) $\rightarrow$ 2 $Cl(3p\ ^2P_{3/2})$.

In the experiments the third harmonic of a Nd:YAG laser (355 nm) was used to photolyse Cl_2, and a Nd:YAG laser pumped dye laser was used for the $(2 + 1)$ REMPI detection of ground state $Cl(^2P_{3/2})$ atoms at 235.336 nm. In such measurements the wavelength of the probe laser radiation was scanned over the Doppler broadened absorption line in order to realize an unbiased detection of all Cl atoms regardless of the v_x component of their velocity. An example of 3-D speed distributions of Cl atoms obtained in this way is shown in Fig. 6.11. Each data point contains information on the velocity components v_x, v_y, v_z. However, due to the restrictions of a 2-D black and white picture, only a projection of this real 3-D distribution can be seen.

The 3-D image of Fig. 6.11 consists of two overlapping spherical distributions: the first (lower) one corresponds to fragments from cold Cl_2 in the molecular beam, and the second (upper) one corresponds to fragments from background Cl_2

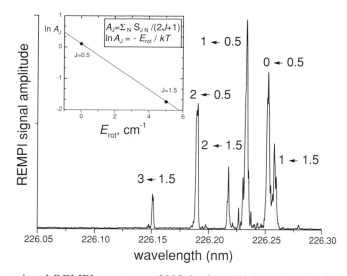

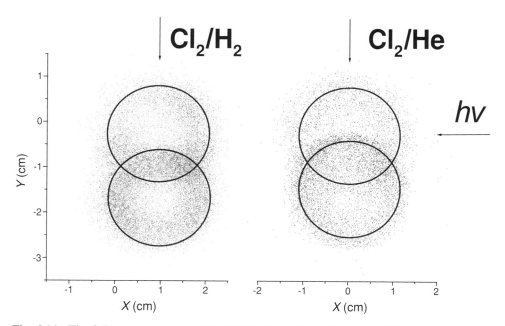

Fig. 6.10. Rotational REMPI spectrum of NO in the cold Ar molecular beam. Only ions from a small area marked by the dotted ellipse in Fig. 6.9 are taken into account. The assignment of the rotational transitions is indicated. In the upper left corner: simplified determination of the rotational temperature T_{rot} of NO. S_{JN} is the peak area for a single $N \leftarrow J$ transition, A_J is the cumulated peak area for all transitions originating from a single J level after correcting for degeneracy.

Fig. 6.11. The 2-D representation of the full 3-D speed distributions of Cl atoms generated in the photolysis of Cl_2 in Cl_2/H_2 and Cl_2/He molecular beams. The lower spheres result from Cl fragments from the photolysis of cold Cl_2 in the molecular beam, whereas the upper spheres result from Cl fragments from thermal background Cl_2. Circles indicate the centres of the Cl fragment speed distributions shown in Fig. 6.13.

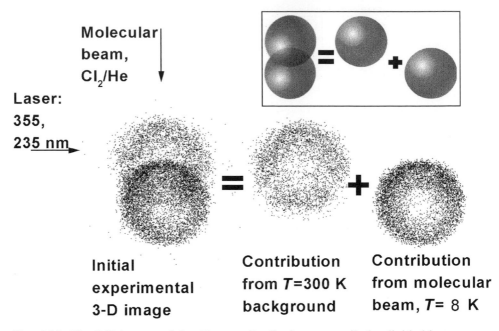

Molecular beam, Cl₂/He

Laser: 355, 235 nm

Initial experimental 3-D image

Contribution from T=300 K background

Contribution from molecular beam, T= 8 K

Fig. 6.12. The 3-D images of the Cl atom distribution can easily be divided in two parts which is not possible with conventional 2-D images. In the upper corner this separation procedure is shown schematically: the experimental 3-D spherical distributions are replaced by half-translucent spheres.

at room temperature. Again, the centres of both distributions are shifted by the distance $L = v_{exp}t$. Thus, from Fig. 6.11 one may easily determine the speed of the molecular beam from the distance between the two rings divided by the TOF. The H_2 beam has been found to be slightly faster than the He beam, see Table 6.1. Now one can, for example, separate these distributions, as is schematically shown in Fig. 6.12, and exclusively study the speed distribution from the dissociation of cold molecules free from background interference. This separation is easily feasible, since for every single event depicted in Fig. 6.11, the knowledge of all three components of the velocity unequivocally determines to which sphere the fragment belongs. This situation is completely different from a conventional 2-D image, because in a 2-D image the spheres occur as two circles and within the common area of these two circles it is impossible to know which data point belongs to which distribution.

The result of our 3-D Cl_2 study is shown in Fig. 6.13. In principle, the Cl fragment speed distribution, hence also the distribution of its kinetic energy E_{kin}, should be single-valued: $E_{kin}(Cl) = \frac{1}{2}[h\nu - D_0(Cl)]$, where the (small) internal energy of the parent molecule has been neglected. Here, $h\nu$ is the energy of the dissociating photon, and D_0 is the dissociation energy for the atomization of the chlorine

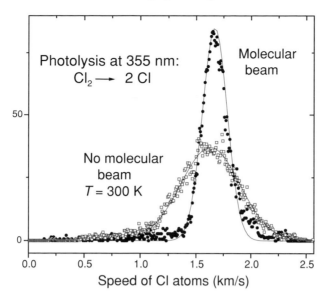

Fig. 6.13. Speed distributions of Cl fragments from the photodissociation of Cl_2 obtained from 3-D images (see Figs. 6.11, 6.12) for a Cl_2/He molecular beam.

molecule. However, this very small thermal energy after cooling of the parent in the supersonic expansion in conjunction with the response function of the apparatus leads to a small broadening of the speed distribution as depicted in Fig. 6.13. We determined not only the average width of the speed distribution of Cl, but also the widths along all three axes. It was found that these widths are different and exhibit a different dependence on the acceleration voltage. The main anisotropic factors, which contribute to the width of the distribution, are the length of the laser spot, rotation of Cl_2, and the speed dispersion of Cl_2 in the molecular beam. From Fig. 6.13 it can be seen that also the widths of the distributions for room-temperature and cold Cl_2 parent molecules are different because of the different speed dispersions and rotational energies.

6.4.2 Spatial fragment anisotropy – how to calculate the β parameter

From the observed 3-D distributions not only the speed distribution can be extracted as demonstrated in the previous section, but it is also possible to determine the angular distribution of the photofragments for any given speed value v. Before specific examples are dealt with, the general approach to studying angular distributions of photodissociation products will be considered. The normalized angular distribution of photofragments with a single given speed with respect to the polarization vector **E** of the electromagnetic field of the photolysis laser is, as we are now familiar, written as

$$P(\theta, \varphi) = \frac{1}{4\pi}[1 + \beta P_2(\cos \theta)]. \tag{6.6}$$

Here θ and φ are the polar and the azimuthal angle of the photofragment recoil velocity v with the polarization vector $\mathbf{E}$, β is the anisotropy parameter which characterizes the angular distribution, and $P_2(x)$ is the second-order Legendre polynomial: $P_2(x) = \frac{1}{2}(3x^2 - 1)$. $P(\theta, \varphi)$ is normalized such that

$$\int_0^\pi \sin \theta d\theta \int_0^{2\pi} d\varphi \, P(\theta, \varphi) = 1.$$

If the integration over φ is performed, then a distribution $P_{th}(\theta)$ with respect to the polar angle may be defined:

$$P_{th}(\theta) = \frac{1}{2} \sin \theta [1 + \beta P_2(\cos \theta)] \tag{6.7}$$

such that

$$\int_0^\pi P_{th}(\theta) d\theta = 1.$$

The β parameter ranges from -1 to $+2$ with an isotropic distribution of the fragments corresponding to $\beta = 0$. Most often, the β parameter is considered to be independent of the fragment velocity. However, this need not be the case. Especially, the broader the kinetic energy distribution becomes, the more important becomes the velocity dependence of β, and the more information on the dynamics of a reaction is contained in this velocity dependence. An interesting application of the 3-D technique is the study of this dependence. This means that we must calculate the β parameter for a group of photofragments that have a velocity within the range $v_0 < v < v_1$. The experimental distribution of the photofragments over the angle θ is given by the function $P_{exp}(\theta)$. The β parameter may be determined from the condition

$$\frac{\partial}{\partial \beta} \int_0^\pi [P_{exp}(\theta) - P_{th}(\theta)]^2 d\theta = 0. \tag{6.8}$$

The normalized experimental distribution of the photofragments over the angle θ may be calculated as

$$P_{exp}(\theta) = \frac{1}{N} \sum_{i=1}^N \delta(\theta - \theta_i), \tag{6.9}$$

where $\delta(x)$ is a delta-function, θ_i is the polar angle for the i-th photofragment, and the summation is performed over particles with velocities in the range of consideration. N is the number of these particles. Inserting (6.9) into (6.8) gives the final result

$$\beta = \tfrac{4}{5}\,(32B_1/\pi + 1) \tag{6.10}$$

where $B_j = \frac{1}{N}\sum_i \sin\theta_i[P_2(\cos\theta_i)]^j$.

The β parameter can also be obtained from a subset of the observed photofragments if the range of the angle θ is accordingly constrained. For example, in the $\mathbf{E}\parallel\mathbf{Z}$ case a contribution from ^{37}Cl atoms to the distribution of ^{35}Cl atoms occurs at some range of the angle θ. Usually, we want to get rid of this contribution. Let us determine the β parameter for a group of the photofragments that are situated in the range $a \le \theta \le b$. The same approach gives the final result

$$\beta = -\frac{A_0^2 B_1 - A_0(A_1 B_0 + C_1) + A_1 C_0}{A_0(A_1 B_1 - C_2) - A_1(A_1 B_0 - C_1)} \tag{6.11}$$

where $A_j = \int_a^b [P_2(\cos\theta)]^j \sin\theta\, d\theta$ and $C_j = \int_a^b [P_2(\cos\theta)]^j \sin^2\theta\, d\theta$. Applying (6.10) or (6.11) to experimental results one can find the $\beta(v)$ dependence.

Concerning the photolysis of Cl_2, at 355 nm the perpendicular optical transition $^1\Pi_{1u} \leftarrow {}^1\Sigma_{0_g^+}$ contributes to more than 90% of the light absorption [22–24]. The excited state is repulsive, so the dissociation is fast and the β parameter must be close to its limiting value of -1. The value determined from our experimental data is $\beta = -1.00 \pm 0.05$ and agrees excellently with previous measurements and theoretical calculations [22].

6.5 Beyond the classic application: modern developments

6.5.1 Atomic photofragment alignment

The strength of the 3-D imaging method lies in the straightforward investigation and analysis of processes where the cylindrical symmetry of the fragment distribution is broken. One such example is the photofragment angular momentum polarization (orientation and alignment) in a photodissociation process. Fragments produced by (circularly or linearly) polarized photolysis may be oriented and/or aligned, that means that they exhibit nonequilibrium populations of the m_J and $|m_J|$ magnetic sublevels, respectively. Measurements of the angular momentum polarization as a function of the recoil velocity vector ($\mathbf{v}$–$\mathbf{J}$ correlation) provide the most detailed information on the dissociation dynamics. While this property has been studied for rotating diatomic photofragments for over a decade, the atomic $\mathbf{v}$–$\mathbf{J}$ correlation has been observed only recently [25–29]. Note that in most essential details, there is little similarity between atomic and molecular photofragment phenomena.

We consider only the case of monoenergetic atomic photofragments produced by linearly or circularly polarized photolysis radiation, and detected by a linearly or circularly polarized probe laser via $(2 + 1)$ REMPI. In the case of linearly polarized photolysis and probe laser radiation, polarization geometries may be realized where the polarization directions are oriented parallel or perpendicular to each other. The angular distribution of the atoms is then given by [28]

$$I_{\parallel}(\theta, \varphi) = \left[1 + \frac{\beta}{2}(3x - 1)\right] + V(a_1 + a_2 x + a_3 x^2) \tag{6.12}$$

$$I_{\perp}(\theta, \varphi) = \left[1 + \frac{\beta}{2}(3x - 1)\right] - \frac{V}{2}(a_1 + a_2 x + a_3 x^2)$$

$$- \frac{3V}{4}(1 - x)\left(a_4 + \frac{2}{3}a_1 x\right)\cos 2\varphi \tag{6.13}$$

where $V = \frac{5R_2}{R_0}\sqrt{\frac{J(J+1)}{(2J+3)(2J-1)}}$ and $x = \cos^2\theta$. θ and φ are the spherical angles with respect to the dissociation laser polarization, the direction of the dissociation laser beam is $\theta = 90°$, $\varphi = 0°$. The β and the a_i are parameters that may be extracted from experiment by fitting the distributions from (6.12) and (6.13) to the experimentally observed data. The a_i parameters are connected to dynamical parameters characterizing the dissociation process by (6.14). J is the electronic angular momentum of the initial state of atoms, which is assumed to be small ($J < 2$). Otherwise the expressions for the distributions are more complex. The alignment sensitivity parameter R_2/R_0 depends on the transition involved to detect the atoms. For example, for $(2 + 1)$ REMPI detection of the ground state $Cl(^2P_{3/2})$ atoms via the intermediate $^2S_{1/2}$ and $^2D_{5/2}$ states (236.284 and 234.640 nm, respectively) the parameter is equal to -1 and $5/7$, respectively [28].

Using the expressions

$$\begin{pmatrix} a_1 \\ \frac{1}{3}a_2 \\ \frac{1}{3}a_3 \\ a_4 \end{pmatrix} = \begin{pmatrix} -1 & -1 & 0 & -3 \\ 1 & 2 & -1 & 2 \\ 0 & -3 & 1 & -1 \\ 1 & 1 & 0 & -1 \end{pmatrix} \begin{pmatrix} \frac{1}{2}s_2 \\ \frac{1}{2}\alpha_2 \\ \gamma_2 \\ \frac{1}{4}\eta_2 \end{pmatrix} \tag{6.14}$$

one can obtain the s_2, α_2, γ_2, and η_2 alignment anisotropy parameters, which are responsible for different excitation mechanisms of the parent molecule.

Consider briefly, for example, the photolysis of Cl_2 at 355 nm [28]. The excited states $C^1\Pi_{1u}$ and $A^3\Pi_{1u}$ can be optically excited from the ground state $X^1\Sigma_{0_g^+}$ of the molecule, the contribution to the light absorption is $> 90\%$ and $< 10\%$, respectively. The photodissociation via $A^3\Pi_{1u}$ state does not produce alignment of Cl atoms, consequently the alignment occurs only via the $C^1\Pi_{1u}$ state. For the photodissociation of Cl_2 by linearly polarized radiation near 355 nm theory predicts the relations

$$\alpha_2 = \tfrac{1}{2}s_2 = -\tfrac{2}{25}(1 - w), \quad \eta_2 = -\tfrac{4\sqrt{6}}{25}w_{\mathrm{coh}}, \quad \gamma_2 = 0, \tag{6.15}$$

where w is the probability of a nonadiabatic transition from $C^1\Pi_{1u}$ to $A^3\Pi_{1u}$ and w_{coh} is the 'coherent' off-diagonal nonadiabatic transition matrix element. In other words, $w = |< C|V|A >|^2$ and $w_{\mathrm{coh}} = \mathrm{Re}[< C|V|A >< C|V|C >^*]^2$. Here V is the operator responsible for the non-adiabatic transition, $< C|$ and $< A|$ denote excited electronic states, and $\Omega = 1$ is assumed in all cases. The experimental values for the alignment anisotropy parameters may be rather well reproduced by (6.15) with $w = 0.60 \pm 0.04$ and $w_{\mathrm{coh}} = -0.19 \pm 0.05$.

The 3-D imaging technique is the most appropriate method for this kind of investigation, since the angular distributions $I_{\parallel}(\theta, \varphi)$ and $I_{\perp}(\theta, \varphi)$ are measured directly. Otherwise the 3-D distributions must be reconstructed from the 2-D images, as was done in the above-mentioned study of photodissociation of Cl_2.

6.5.2 Application of 3-D detection to photodissociation processes

In the study of photodissociation dynamics the experimental goal is simple in concept: to detect and characterize the products resulting from photon absorption, and to determine the effect of photon energy and polarization on the nature and distribution of these products. Until recently the angular distribution of the products was generally characterized by a single, velocity-independent β parameter (6.6). Now the combination of molecular beam techniques with REMPI/TOF mass spectrometry allows us to obtain more detailed information, for example, the dependence of the β parameter on the particle velocity v.

6.5.3 Angular distribution of photofragments: 'core-sampling' technique

In principle, the determination of the $\beta(v)$ dependence does not require 3-D or 2-D imaging techniques. Let us briefly consider an experiment performed by K. Liu and co-workers [30] in which the photodissociation of NO_2 at 355 nm was studied:

$$NO_2 + h\nu \rightarrow NO\big(^2\Pi_\Omega\big) + O\big(^3P_J\big).$$

Oxygen atoms in all three spin-orbit states $O(^3P_J)$ ($J = 2, 1, 0$) were state-selectively ionized by a $(2 + 1)$ REMPI detection scheme around 226 nm [31]. The TOF spectra of the $O(^3P_J)$ atoms were monitored by the 'core-sampling' technique which detects only the 'core distribution', i.e., those ions with very small v_x and v_y components. The selection of the ions with $v_x \approx v_y \approx 0$ was achieved by placing a small aperture in front of the ion detector. Generally, the photofragment centre of mass velocity distribution over speed v and angle θ can be expressed as

$$P(v, \theta) = g(v)[1 + \beta(v)P_2(\cos \theta)], \tag{6.16}$$

where the speed distribution $g(v)$ and $\beta(v)$ are unknown functions. Normalization factors are included in the $g(v)$ dependence. Then the TOF spectra for both polarization directions of the photolysis radiation may be expressed as

$$\theta = 0° \quad \mathbf{E} \,\|\, \mathbf{Z} \quad P_{\|}(v) = g(v)[1 + \beta(v)]$$
$$\theta = 90° \quad \mathbf{E} \perp \mathbf{Z} \quad P_{\perp}(v) = g(v)[1 - {}^1/_2\beta(v)]$$

where $\mathbf{E}$ is the electric field vector of the photolysis radiation. After normalization, the functions $P_{\|}$ and $P_{\perp}$ may be used to obtain $\beta(v)$:

$$\beta(v) = 2\frac{P_{\|}(v) - P_{\perp}(v)}{P_{\|}(v) + 2P_{\perp}(v)}. \tag{6.17}$$

This expression was applied to the experimental data and $\beta(v)$ was determined for all three states of $O(^3P_J)$ (Fig. 6.14). The β distributions for all three sublevels of the $O(^3P_J)$ state display a downward trend with the increase in rotational, vibrational, and spin-orbit energy of NO. The explanation of this trend seems to be quite straightforward: if the photon energy is redistributed over vibrations of NO_2, then the $O(^3P_J)$ atoms are distributed over a wider range of angles. Consequently, β becomes smaller, the rotational and vibrational excitation of NO become higher and the kinetic energies of NO and $O(^3P_J)$ become smaller.

It should be noted, however, that whereas 3-D imaging by a DLD is universally applicable regardless of the specific form of the spatial fragment distribution, core sampling is in exactly the same way limited to the investigation of cylindrically symmetric fragment distributions as photofragment imaging and velocity mapping. Basically, the simultaneous measurement of two dimensions by a PSD is substituted by a sequence of two one-dimensional measurements with identical information content.

6.5.4 Application of 3-D detection to the photolysis of $COCl_2$

The 3-D photofragment imaging technique is a much more powerful and general method than the core-sampling technique discussed above. First, the 3-D imaging technique deals with much more detailed information because the complete 3-D velocity distribution is obtained directly. Second, background contributions from room-temperature molecules or from other ions with the same mass-to-charge ratio may easily be removed. Finally, 3-D imaging does not deal with the profile of

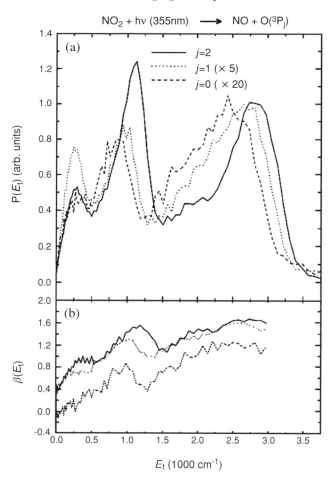

Fig. 6.14. (a) Photofragment centre of mass translational energy distributions from the photodissociation of NO_2 at 355 nm [30]. The results, normalized by the relative populations 1:0.21:0.05, for the $J = 2$, 1, and 0 fine-structure states of $O(^3P_J)$ are represented by solid, dotted, and dashed lines, respectively. (b) The dependence of the β parameters on the Cl fragment translational energy for the three states of $O(^3P_J)$.

an analogue signal which may be sensitive to the apparatus broadening function, saturation of the MCP assembly etc.

A very promising subject for the application of 3-D photofragment imaging techniques is the photoinduced three-body decay

$$ABC \rightarrow A + B + C,$$

where A, B and C may be atoms, diatomics or polyatomics. The three-body processes are important since they govern complex chemical reactions including combustion processes and atmospheric chemistry, and they are much more complicated in comparison with common two-body decays. The first question aims at the mechanism of a three-body decay, which may be sequential, synchronously concerted, or asynchronously concerted. We call the three-body decay sequential if the time between the first step, ABC → AB + C, and the second step, AB → A + B, is greater than the mean rotational period of the AB fragment. Otherwise, the decay is considered to be concerted. A synchronously concerted mechanism assumes the linear momenta of A and C fragments to be equal, whereas an asynchronously concerted mechanism does not. A very illustrative, yet oversimplifying picture of synchronously and asynchronously concerted mechanisms is decays via symmetric and antisymmetric stretching modes of a linear ABA parent molecule, respectively. The mechanism of the decay can be understood entirely by examining the spatially and temporally resolved photodissociation products.

A very good example suited to illustrate the subject is the photodissociation of phosgene:

$$COCl_2 + h\nu \rightarrow CO + 2Cl \qquad D_0(Cl\text{–}CO\text{–}Cl) = 3.56\,eV$$

where $E_{av} = h\nu - D_0 = 1.716$ eV is the energy available to the products resulting from 235 nm photolysis. It is well known that in this process two-body decay channels (Cl + COCl, CO + Cl_2) are negligible [32,33]. In our experiments the Cl atoms in both spin-orbit states $Cl(^2P_J)$ (J = 3/2, 1/2) were state-selectively ionized by a (2 + 1) REMPI detection scheme around 235 nm [34]. The 3-D velocity distributions were obtained according to the procedure described in the previous section. The speed distribution and the $\beta(v)$ dependence for ground state $Cl(^2P_{3/2})$ are shown in Fig. 6.15. This is a rather simple picture, which can easily be interpreted. From Fig. 6.15 it can be seen that the velocity distribution of Cl atoms consists of two peaks. For one of them the β parameter is close to 0 and for the other one the β parameter is about 0.5.

Evidently, the fast Cl atoms are produced in the first step, $COCl_2 \rightarrow COCl^* + Cl$, because their angular distribution is anisotropic. The slow Cl atoms are produced in the second step, $COCl^* \rightarrow CO + Cl$, which produces an almost isotropic angular distribution of Cl atoms. This means that the decay of $COCl_2$ is asynchronously concerted, or sequential, with a rather large time between the first and the second steps. This description of the decay of $COCl_2$ is too simple, however, since the distributions for the ground state $Cl(^2P_{3/2})$ and spin-orbit excited state $Cl(^2P_{1/2})$ are different from each other. According to a more detailed analysis, about 80% of the decay correspond to the asynchronously concerted mechanism and about 20% to the synchronously concerted mechanism [32,33].

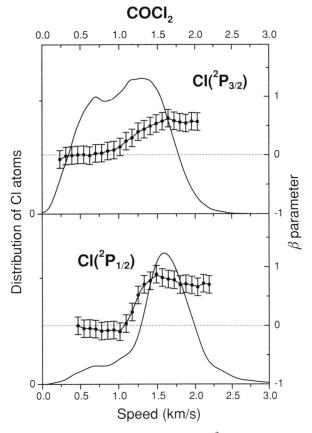

Fig. 6.15. Velocity distributions for the ground state $Cl(^2P_{3/2})$ and excited $Cl(^2P_{1/2})$ atoms in the photolysis of $COCl_2$. The dependence of the β parameters on the Cl fragment velocity (right scale) is shown by curves with error bars.

6.5.5 Photolysis of SOCl₂: 3-D detection of Cl atoms

Another example of a three-body decay studied by the 3-D imaging technique is the photodissociation of $SOCl_2$ at 235 nm. At this wavelength there are two energetically possible channels that give Cl atoms:

$$SOCl_2 + h\nu \rightarrow SO + 2\,Cl,\ SOCl + Cl.$$

The speed distribution and $\beta(v)$ dependence for ground state $Cl(^2P_{3/2})$ atoms are similar to those for the dissociation of $COCl_2$ (Fig. 6.16). From Fig. 6.16 it can be seen that the Cl atoms may be separated into two groups, fast and slow. The angular distribution of the fast atoms is rather anisotropic and hence they are produced in the first step, $SOCl_2 \rightarrow SOCl^* + Cl$, while the slow Cl atoms are produced

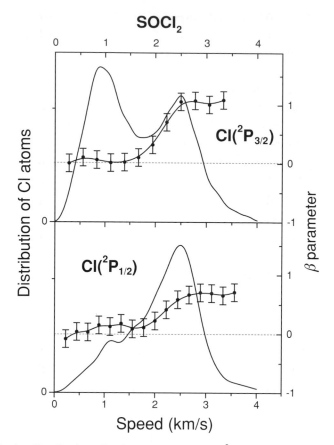

Fig. 6.16. Velocity distributions for the ground state $Cl(^2P_{3/2})$ and excited $Cl(^2P_{1/2})$ atoms in the photolysis of $SOCl_2$. The dependence of the β parameters on the Cl fragment velocity (right scale) are shown by curves with error bars.

in the second step, $SOCl^* \rightarrow SO + Cl$, described by an almost isotropic angular distribution of Cl atoms. Thus the three-body decay of $SOCl_2$ is asynchronously concerted or sequential, but not synchronously concerted.

6.6 The road ahead and concluding remarks

Three-dimensional photofragment imaging techniques have proved to be a powerful tool to study the velocity (speed and angle) distributions of the products of photodissociation processes. The progress is directly comparable with real 3-D (stereo) pictures relative to 2-D photos. It is hoped that the abundance of new experimental information available for this new method will stimulate future theoretical studies. Generally, the merits of this method lie in the investigation of photochemical processes where no cylindrical symmetry with respect to any axis is maintained, covering all aspects of reactant and product alignment and orientation. Here, 3-D imaging allows the most direct measurements of the underlying effects, regardless of whether they are induced in the preparation of the sample or by the dynamics of the elementary reaction.

Further extensions of this method will probably go in two directions: first, there are coincidence measurements, that is, the simultaneous observation of two fragments resulting from photodissociation of a parent molecule. This technique must directly provide all three speed components of both the fragments. Of particular interest may be applications of this technique to study the photoinduced three-body dissociation (ABC → A + B + C) since for these processes only a combination of 3-D imaging with coincidence measurements provides enough information for a complete kinematic analysis.

Other directions of future research may be the detailed study of speed distributions of the products of chemical reactions or from the photodissociation processes with vibrational preexcitation of the reactants.

Acknowledgements

We appreciate the great support of Dr. U. Titt and Dr. M. Roth. The authors A. Chichinin and T. Einfeld gratefully acknowledge the support of the Alexander von Humboldt Foundation and the Fonds der chemischen Industrie.

References

1. A. J. R. Heck, D. W. Chandler, *Ann. Rev. Phys. Chem.* **46**, 335–72 (1995).
2. V. Aquilanti, D. Ascenzi, M. de Castro Vítores, F. Pirani, D. Cappelletti, *J. Chem. Phys.* **111**, 2620–32 (1999).
3. F. Pirani, D. Cappelletti, M. Bartolomei, V. Aquilanti, M. Scotoni, M. Vescovi, *et al.*, *Phys. Rev. Lett.* **86**, 5035–8 (2001).
4. V. Aquilanti, D. Ascenzi, D. Cappelletti, F. Pirani, *Nature* **371**, 399–402 (1994).
5. V. Aquilanti, D. Ascenzi, D. Cappelletti, F. Pirani, *J. Phys. Chem.* **99**, 13620–6 (1995).
6. V. Aquilanti, D. Ascenzi, D. Cappelletti, R. Fedeli, F. Pirani, *J. Phys. Chem. A* **101**, 7648–56 (1997).
7. S. Harich, A. M. Wodtke, *J. Chem. Phys.* **107**, 5983–6 (1997).
8. E. B. Anthony, W. Schade, M. J. Bastian, V. M. Bierbaum, S. R. Leone, *J. Chem. Phys.* **106**, 5413–22 (1997).
9. D. P. de Bruijn, J. Los, *Rev. Sci. Instrum.* **53**, 1020–6 (1982).
10. J. C. Brenot, M. Durup-Ferguson. In: *State-Selected and State-to-State Ion-Molecule Reaction Dynamics*, Part 1: Experiment, ed. C. Y. Ng, M. Baer, (Wiley, New York, 1992), pp. 309–99.
11. H. Helm, P. C. Cosby, *J. Chem. Phys.* **86**, 6813–6822 (1987).
12. R. E. Continetti, D. R. Cyr, D. L. Osborn, D. J. Leahy, D. M. Neumark, *J. Chem. Phys.* **99**, 2616–31 (1993).
13. M. Takahashi, J. P. Cave, J. H. D. Eland, *Rev. Sci. Instrum.* **71**, 1337–44 (2000).

14. C. Maul, K.-H. Gericke, *Int. Rev. Phys. Chem.* **16**, 1–79 (1997).
15. C. Maul, K.-H. Gericke, *J. Phys. Chem. A* **104**, 2531–41 (2000).
16. J. Danielak, U. Domin, R. Kepa, M. Rytel, M. Zachwieja, *J. Mol. Spect.* **181**, 394–402 (1997).
17. D. R. Miller, In: *Atomic and Molecular Beam Methods* vol. 2, ed. G. Scoles, (Oxford University Press, New York, Oxford, 1992), pp. 14–53.
18. S. E. Sobottka, M. B. Williams, *IEEE Trans. Nucl. Sci.* **35**, 348–51 (1988).
19. O. Jagutzki, V. Mergel, K. Ullmann-Pfleger, L. Spielberger, U. Dörner, H. Schmidt-Böcking, *Nucl. Instrum. Methods A*, **447**, 244–9 (2002).
20. M. Lampton, O. Siegmund, R. Raffanti, *Rev. Sci. Instrum.* **58**, 2298–2305 (1987).
21. Z. Amitay, D. Zajfman, *Rev. Sci. Instrum.* **68**, 1387–92 (1997).
22. T. Ishiwata, A. Ishiguro, K. Obi, *J. Mol. Spectrosc.* **147**, 300–20 (1991).
23. Y. Matsumi, K. Tonokura, M. Kawasaki, *J. Chem. Phys.* **97**, 1065–71 (1992).
24. P. C. Samartzis, I. Sakellariou, T. Gougousi, T. N. Kitsopoulos, *J. Chem. Phys.* **107**, 43–8 (1997).
25. A. J. Alexander, Z. H. Kim, S. A. Kandel, R. N. Zare, T. P. Rakitzis, Y. Asano, S. Yabushita, *J. Chem. Phys.* **113**, 9022–31 (2000).
26. Y. Mo, T. Suzuki, *J. Chem. Phys.* **112**, 3463–73 (2000).
27. A. S. Bracker, E. R. Wouters, A. G. Suits, Y. T. Lee, O. S. Vasyutinskii, *Phys. Rev. Lett.* **80**, 1626–29 (1998).
28. A. S. Bracker, E. R. Wouters, A. G. Suits, O. S. Vasyutinskii, *J. Chem. Phys.* **110**, 6749–65 (1999).
29. K. O. Korovin, B. V. Picheyev, O. S. Vasyutinskii, H. Valipour, D. Zimmerman, *J. Chem. Phys.* **112**, 2059–62 (2000).
30. C.-H. Hsieh, Y.-S. Lee, A. Fujii, S.-H. Lee, K. P. Liu, *Chem. Phys. Lett.* **277**, 33–8 (1997).
31. D. J. Bamford, M. J. Dyer, W. K. Bischel, *Phys. Rev. A* **36**, 3497–3500 (1987).
32. C. Maul, T. Haas, K.-H. Gericke, F. J. Comes, *J. Chem. Phys.* **102**, 3238–47 (1995).
33. C. Maul, T. Haas, and K.-H. Gericke, *J. Phys. Chem. A* **101**, 6619–32 (1997).
34. S. Arepalli, N. Presser, D. Robie, R. J. Gordon, *Chem. Phys. Lett.* **118**, 88–92 (1985).

7

Photoelectron and photoion imaging with femtosecond pump-probe time clocking

TOSHINORI SUZUKI AND BENJAMIN J. WHITAKER

7.1 Introduction

One of the most exciting advances in chemical physics in recent years has been the emergence and development of femtochemistry. This has been brought about largely because of advances in ultrafast laser technology, particularly the discovery of self-mode locking in Ti:sapphire and the development of chirped pulse or regenerative amplifiers. Another important innovation has been the development of a variety of linear and nonlinear spectroscopic techniques to probe electronic and nuclear dynamics. Nonlinear methods have been particularly useful in the study of solvation dynamics in the condensed phase. In the gas phase, where the density of molecules is much lower, ionization techniques such as pump-probe mass spectrometry have more often been employed. However, mass spectrometry can only provide the time-dependent population of a chemical species, in other words, kinetic information. In order to extract more detailed information on the reaction dynamics, measurements of the velocity vectors of the photoelectrons and fragment ions produced upon ionization are required. As we have seen in the preceding chapters, an imaging detector placed at the end of a time-of-flight mass spectrometer can easily accomplish such measurements. In this chapter we explore how ultrafast lasers can be coupled with charged particle imaging to develop experimental probes of ultrafast dynamic processes in molecules, such as electronic dephasing (radiationless transitions) and intramolecular vibration energy redistribution (IVR).

7.2 Femtosecond lasers

Ultrashort pulses are generated by mode-locked lasers. For a laser to mode-lock it must be made to oscillate at a large number of frequencies (the longitudinal modes supported by the cavity) and these oscillations must be brought into phase with one another. Various techniques have been employed to achieve this result.

In active mode-locking, a modulator is placed in the cavity which chops the light field at a frequency corresponding to the round trip time of a photon within the resonator. In passive mode-locking systems, a saturable absorber is used so that as the field in the cavity builds up beyond a certain threshold the absorbing 'shutter' becomes transparent. In the early 1990s a new scheme called self-mode-locking was developed [1] and has become by far the most common method for the generation of ultrashort optical pulses. The phenomenon was first discovered in titanium doped sapphire (Ti:sapphire), although other materials have since been found to exhibit the same effect. Ti:sapphire has a gain bandwidth from about 700 to 1100 nm, peaking close to 800 nm, and has the broadest gain curve of any of the solid-state lasing media yet discovered. The broad gain curve is important for the generation of ultrashort pulses because, as a consequence of the Fourier transform lifetime broadening relationship

$$\tau \delta E \geq \frac{h}{2}, \tag{7.1}$$

the spectral density function of an exponentially decaying excited state is a Lorentzian function whose width, δE (J), is inversely proportional to the decay time, τ (s). The lifetime broadening relation is obviously similar to the Heisenberg uncertainty relationship, although its origins are somewhat different, since time is not a quantum mechanical operator [2]. The consequence of (7.1) is that short pulses are necessarily spectrally broad. For example, a transform-limited 10 fs pulse (full width at half-maximum) centred at 800 nm has a bandwidth of close to 100 nm. This means that in order to get short pulses the laser must exhibit gain over a wide spectral range and that the mirror coatings etc. in the cavity must equally cover a similarly broad range. In practice, the transform limit is never achieved because of spectral imperfections. A summary of the time-bandwidth properties of the more common model pulse shape functions is given in Table 7.1.

In self-mode-locking the longitudinal cavity modes are held in phase through the action of an instantaneous Kerr lens in the lasing medium. In a typical Ti:sapphire oscillator the peak fluence approaches 10^9 W cm^{-2}. At such high intensities the refractive index of the Ti:sapphire crystal, the laser rod, in the oscillator becomes a function of the incident intensity through the optical Kerr effect:

$$n(I) = n_0 + n_1 I + n_2 I^2 \cdots \tag{7.2}$$

where n_0 is the normal refractive index of the material and n_1 is the first nonlinear correction, etc. The intensity profile of the laser across the lasing medium is determined by the transverse mode structure of the cavity, and the transverse mode with the smallest diffraction loss, and hence the highest gain, has a Gaussian profile. The refractive index gradient induced in the medium by the optical Kerr effect

Table 7.1. *Time-bandwidth products for various pulse shapes [3], where τ_p = pulse width (fwhm) and Δv = spectral width (fwhm).*

$I(t)\ (x \equiv t/T)$	$\Delta v \tau_p$	τ_p/T
Gaussian $I(t) = \exp(-x^2)$	0.4413	$2\sqrt{\ln 2}$
Diffraction function $I(t) = \dfrac{\sin^2 x}{x^2}$	0.8859	2.7831
Hyperbolic sech $I(t) = \text{sech}^2 x$	0.3148	1.7627
Lorentzian $I(t) = \dfrac{1}{1+x^2}$	0.2206	2

is, therefore, strongest at the centre of the profile and so creates a positive lens that brings the light to a focus or, for a divergent beam, restores its collimation.

The laser medium (rod) of the Ti:sapphire laser is collinearly pumped by a continuous wave (CW) laser, usually a diode-pumped neodymium laser with intracavity frequency doubling. The pump beam is focused to a waist, which is smaller than the transverse mode profile of the cavity. The Ti:sapphire laser then begins to oscillate on a large number of longitudinal modes as a CW laser. Under these conditions the phase relationship between the modes fluctuates randomly. However, noise in the system may cause some of the longitudinal modes to accidentally come into phase. The more intense field in the noise pulse, in which some of the longitudinal modes are in phase, causes self-focusing within the crystal and makes the beam overlap more with the pumped volume, until complete overlap between the pulse and the pumped volume saturates the gain, at which point the oscillator self-mode-locks and produces a train of short pulses. To establish the pulsing one has to introduce a noise spike. Stylishly, by tapping one of the end mirrors of the cavity with one's finger!

Because light at different wavelengths travels at different speeds through a dispersive medium, the component frequencies of the pulse are separated in time as they pass through the Ti:sapphire rod. There are two points to note here. First the centre of the pulse is delayed with respect to a pulse travelling in air. This is called the group delay and is not a broadening effect. Second, normally dispersive media, such as the Ti:sapphire rod or glass, impose a positive frequency sweep or 'chirp' on the pulse, meaning that the blue components are delayed with respect to the red. This effect is called group velocity dispersion, and is obviously undesirable for the generation of ultrashort pulses since it will cause the laser pulse to be broadened each time it passes through the gain medium. The effect can be compensated for by inserting a pair of prisms into the cavity, as indicated in Fig. 7.1. This arrangement (see also Fig. 7.2a) creates a longer path through the prism material (usually fused

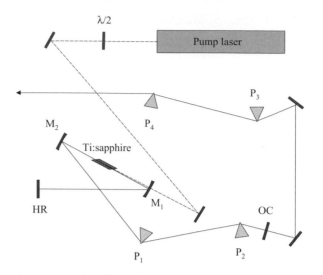

Fig. 7.1. Basic cavity design of a self-mode-locked ultrafast oscillator. The cavity is formed by a high reflector, HR, and a $\sim 10\%$ output coupler, OC. The Ti:sapphire gain medium is positioned between two fold mirrors, M_1 and M_2, and is collinearly pumped through one of the folding mirrors by ~ 5 W cw blue/green light from an Ar-ion or diode pumped Nd:YAG/Nd:YLF laser. The prisms P_1 and P_2 within the cavity control the spectral dispersion (chirp) introduced in the laser rod. The extracavity prisms P_3 and P_4 undo the chirp introduced by P_1 and P_2 in the final pass through the cavity. The Ti:sapphire rod is Brewster angle cut to minimize losses in the cavity and is typically 5 mm in length doped to absorb about 90% of the incident pump light.

silica) for the red wavelengths compared with the blue and so introduces a compensating negative dispersion. In the oscillator design shown in Fig. 7.1, the pulse is coupled out of the cavity from the dispersed end of the cavity and a second pair of matching extracavity prisms is required to undo the negative chirp on the final pass. An alternative solution to the use of prisms to compensate for the dispersion in the rod can be obtained with special mirror coatings which have dispersion opposite to that of the rod, effectively reflecting the blue light from the top surfaces of the mirror and the red light from dielectric coatings deeper in the mirror so that the red light travels a greater distance in the cavity than the blue [4].

Control of dispersive frequency chirp is essential in order to deliver a short pulse to the sample since the pulse is generally lengthened as it passes through various optical elements on its way to the sample. It is also the basis of chirped-pulse amplification.

The pulse energy from a self-mode-locked Ti:sapphire oscillator is usually at the nanojoule level. Although the peak power in such a pulse is high because of its short duration, the photon flux in the pulse is often not sufficient for many of the experiments we might want to do. However, amplification of these pulses to the millijoule level and above is complicated by the extremely high peak powers

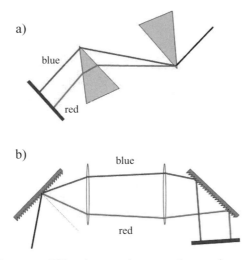

Fig. 7.2. Pairs of prisms or diffraction gratings can be used to control the dispersion. The paths of two rays of different wavelength (blue and red). The arrangement shown in (a) creates a longer optical path for the longer wavelength light compared to the shorter wavelength, introducing negative dispersion when the light is reflected back through the prism pair. The arrangement shown in (b) creates a longer path for the shorter wavelength light compared to the longer wavelength, introducing positive dispersion when the light is reflected back through the grating pair. Positive dispersion has the effect of stretching the input pulse in time.

involved. For example, a 1 mJ 100 fs pulse focused to a 100 μm spot has a peak fluence of 1.3×10^{14} W cm^{-2}. This is about three orders of magnitude larger than the damage threshold of most optical materials, including the Ti:sapphire crystal in the amplifier! The problem is overcome by stretching the pulse from the oscillator with a normally dispersive element, usually a pair of diffraction gratings as illustrated in Fig. 7.2b, before amplifying it. After amplification the pulse can be recompressed in time by applying a negative chirp with a second matched pair of gratings.

A self-mode-locked oscillator will normally generate a pulse train at between 75 and 85 MHz depending on the length of the laser oscillator; this frequency being the inverse of the round trip time around the oscillator cavity. In order to amplify the pulses, the light from the oscillator must pass through a second slab of the lasing material in which a population inversion has been created. As the pulse passes through the inverted medium its intensity is increased by the stimulated emission of further photons. In order to extract all the energy stored in the amplifying rod the (stretched) pulse from the oscillator is usually sent through the amplifier several times, so that it builds in intensity at each pass, before being switched out of the amplifying cavity. This process is known as regenerative amplification. The injection and extraction of pulses into the amplifier is achieved by polarization switching using a fast Pockels cell, as illustrated in Fig. 7.3. The amplifier rod is

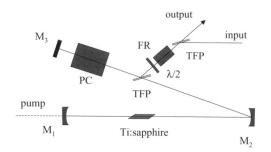

Fig. 7.3. Schematic design of a regenerative amplifier. The arrangement shown is often used for kilohertz operation and other designs may be more appropriate for high-energy low repetition rate operation. The cavity is made up of the three mirrors, M_1, M_2 and M_3 in a folded configuration. The amplifying Ti:sapphire rod is pumped coaxially through the end mirror of the cavity. A Pockels cell (PC) is used to switch in one of the stretched pulses from the Ti:sapphire oscillator synchronously with the laser pulse pumping the amplifier rod. The pulse is allowed to extract gain from the amplifier rod for a number of passes through the cavity, typically ~ 10, before the polarization of PC is switched to eject the amplified pulse from the cavity via the thin film polarizer (TFP). The combination of a Faraday rotator (FR) and the half-wave plate ($\lambda/2$) acts as an optical diode to prevent the amplified pulse feeding back into the oscillator.

usually pumped with a Nd:YAG or Nd:YLF laser operating between 10 and 3000 Hz repetition rate, depending on the application.

Most molecules of interest to us do not absorb light at 800 nm, but the light from a regenerative amplifier is extremely intense and so can be doubled efficiently even in thin samples of nonlinear optical materials (remember that most materials are normally dispersive and impose a positive frequency chirp on a short pulse passing through them, so it is highly desirable to keep the total amount of material between the laser and the experiment as small as possible). Using materials such as $LiNiO_3$ and β-barium borate (BBO) then the fundamental light can be easily doubled, tripled or quadrupled to give pulses centred close to 400, 267 or 200 nm. To generate tunable light optical parametric amplifiers (OPAs) can be employed. These are based on the parametric interaction of a pump field, ω_{pump}, with the molecules in a crystal with sufficient nonlinear susceptibility. The interaction can be thought of as inelastic scattering of the pump photon, $\hbar\omega_{pump}$, whereby the pump photon is absorbed and two new photons, $\hbar\omega_{signal}$ and $\hbar\omega_{idler}$, are generated. Clearly conservation of energy requires that $\omega_{pump} = \omega_{signal} + \omega_{idler}$. The high frequency component is known as the signal and the low frequency component is known as the idler. In a collinear geometry the temporal output of the OPA is limited by the group velocity mismatch between the pump and the generated signal and idler fields. To produce the shortest possible pulses a noncollinear arrangement, as illustrated in Fig. 7.4, is generally employed. In this design, a white light continuum seed

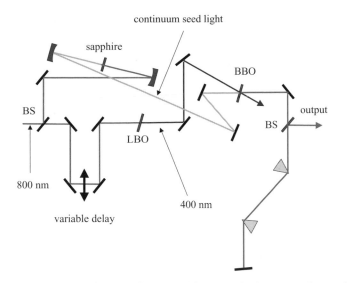

Fig. 7.4. Schematic diagram of a noncollinear optical parametric amplifier.

pulse, generated by stimulated Raman scattering in a thin (1 mm) piece of sapphire, is amplified by the frequency doubled output of a Ti:sapphire amplifier, which provides the pump beam for the OPA. The pump beam is focused onto a BBO crystal to produce a cone of parametric superfluorescence. At a particular angle of incidence, the group velocity of the superfluorescence is practically phase matched for a wide range of frequencies, so that there is no appreciable spatial divergence. By directing the continuum seed beam along the cone axis, a large spectral bandwidth from the white light can be simultaneously phase matched. Adjustment of the relative delay between pump and seed allows one to control the centre wavelength of the amplified light because of the chirp present in the seed pulse. The amplified beam is recompressed using a prism pair yielding sub-20 fs pulses tunable from about 480 to 700 nm or 1000 to 2000 nm depending on whether the idler or signal beam is selected by the final beam splitter. To generate tunable light into the UV or further into IR frequency mixing techniques can be used.

7.3 Time resolved photoelectron imaging: a new probe of femtochemistry

Photoelectron spectroscopy is often introduced in undergraduate courses as supporting evidence for the existence of molecular orbitals, and for small molecules, such as the diatomic molecules formed from the $[..]2p^n$ atoms, the story is convincing. But how 'real' are molecular orbitals in more complicated molecules? More concretely, what can we hope to learn about the dynamics of photochemical processes from photoelectron spectroscopy? Could we, perhaps, use photoelectron spectroscopy to follow the electronic rearrangements during the course of a chemical reaction?

The use of pump-probe photoionization has advantages compared to pump-probe laser induced fluorescence (LIF) for the study of the excited state dynamics since ionization can occur from any part of the excited state potential for a sufficiently energetic photon. The photoabsorption wavelength in a pump-probe LIF experiment using a bound–bound transition between neutral states is a function of nuclear coordinates, so the probe laser frequency needs to be scanned in order to observe the entire region of the potential energy surface. Due to the bound-free nature of ionization, on the other hand, the photoelectrons can carry away the energy difference between the different vibrational levels of the cationic state:

$$E_e = \hbar(\omega_1 + \omega_2) - IP_0 - E_v^+ \qquad (7.3)$$

where E_e is the photoelectron kinetic energy, IP_0 is the adiabatic ionization potential, and E_v^+ is the vibrational energy of the cation. This means that the resonance condition can be fulfilled for ionization of a wave-packet to a range of vibrational levels in the cationic state with a *fixed laser frequency*. This is experimentally easier than the pump-probe LIF experiment and furthermore the charged products can be detected very efficiently.

The photoelectron kinetic energy distribution is determined by the Franck–Condon factors between the vibrational wavepacket in the neutral excited state, $|\psi(t)\rangle$ and the vibrational wavefunctions in the cationic state, $\langle \chi_v^+|$:

$$P(E_e) \propto \sigma \sum_v |\langle \chi_v^+ \mid \psi(t)\rangle|^2 \qquad (7.4)$$

where σ is the photoionization cross-section. If the neutral state changes its electronic character, the vibrational wavepacket and consequently the photoelectron energy distribution will also change, providing the way to probe electronic transition during the course of chemical reaction by pump-probe photoelectron spectroscopy.

The implementation of photoelectron spectroscopy is obviously concerned with the development of techniques that measure the kinetic energy of the ejected electrons when a molecule is ionized. This can be achieved in a number of ways. For (pseudo) continuous light sources, electrostatic energy analysers have been widely utilized, analogously to the magnetic sector mass filters employed in mass spectrometry. When a pulsed laser is used as the ionization source a time-of-flight (TOF) method, in which one measures the arrival time of the photoelectrons at the detector with respect to the laser pulse, is often convenient. Both electric sector and conventional TOF energy spectrometers are inefficient because one can only collect a small fraction of the total photoelectrons, typically much less than 1%. The problem of the small acceptance angle of TOF electron spectrometers can be much improved

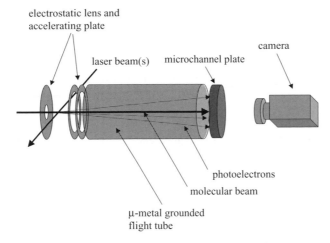

electrostatic lens and
accelerating plate

camera

laser beam(s)   microchannel plate

photoelectrons

molecular beam

µ-metal grounded
flight tube

Fig. 7.5. Schematic diagram of a photoelectron imaging spectrometer.

(to 2π steradian) by use of a magnetic bottle spectrometer [5], but unfortunately at the cost of angular resolution. The measurement of photoelectron angular distribution using electrostatic analysers has been notoriously time-consuming and inaccurate for small acceptance angle detectors due to spatial and temporal variations of the experimental conditions upon rotating the detector angle. Furthermore, in these detection methods the photoelectrons cannot be easily accelerated, and the small mass/charge ratio of electrons makes the experiment extremely sensitive to stray fields, which can only be taken care of with considerable experimental expertise.

Charged particle imaging dramatically improves both sensitivity and angular resolution. The advantage of a photoelectron imaging (PEI) spectrometer, illustrated in Fig. 7.5, is that all the photoelectrons are perfectly collected by an electric field. The detection efficiency is only limited by the quantum yield, about 0.6, of the microchannel plate (MCP) detector. Since PEI measures photoelectrons with different wave-vectors simultaneously, it provides reliable measurements of both speed and angular distributions in a single experimental configuration. In PEI, the photoelectrons are accelerated in order to make the whole electron cloud smaller than the diameter of the position-sensitive detector. Because of the high kinetic energies thus imparted to the photoelectrons, they are practically insensitive to stray fields in and around the apparatus. With velocity mapping and real-time image processing such as event counting, the photoelectron energy resolution reaches the instrumental limit given by the number of pixels in the CCD camera [6]. For a 400×400 pixel camera, the energy resolution for the fastest electron is $v/\Delta v = 140\text{--}200$, or $\Delta E = 10\text{--}14\,\text{meV}$ at $E = 1\,\text{eV}$. PEI also provides uniform sensitivity for photoelectron energies down to extremely slow electrons ($E_{\text{kin}} < 0.2\,\text{eV}$). This is in contrast to magnetic bottle spectrometers that are relatively insensitive to slow photoelectrons.

7.3.1 Examples of time-resolved photoelectron imaging

The electronic spectra of polyatomic molecules can be extremely complicated and difficult to resolve. The electronic spectrum of a polyatomic molecule at room temperature consists of a number of transitions associated with the vibrational and rotational fine structure of the energy levels, as in diatomic molecules, but the spectrum is more complex because of the plethora of vibrational modes and the density of the rotational states, whose spectral signature is superimposed on a particular electronic transition. This often makes the spectrum virtually structureless. However, this is not the complete story, since even if apparent complexity is removed by cooling the sample to very low temperatures (as in a supersonic molecular beam) so that the rovibrational substructure no longer complicates the spectrum, the broad appearance of the absorption spectrum may still remain. This is due to the lifetime broadening of the excited electronic state induced by mixing with lower-lying electronic states. Such mixing occurs, since the Born–Oppenheimer approximation, which assumes the electronic and nuclear motions can be separated, does not hold when two potential energy surfaces closely approach one other or intersect. (If the two electronic states belong to the same symmetry species in a given point group, their potential energy surfaces can come extremely close to each other but never intersect, which is called an avoided crossing of potential energy surfaces.) Although such phenomena also occur in diatomic molecules, in polyatomics they are the rule rather than the exception. The electronic relaxation (or transitions) induced by nuclear motions are generally known as nonadiabatic transitions. Historically, these processes with and without a change of the spin multiplicity of the electronic states are called intersystem crossing (ISC) and internal conversion (IC), respectively.

Figure 7.6a illustrates the intersection of two parabolic potential energy surfaces characteristic of simple two-dimensional harmonic oscillators. In this case the two surfaces meet at a point, since the two states fall into the same symmetry species of the point group except for this singular point. This type of surface crossing is called a conical intersection. In general the topography of the crossing is more complex than this and the surfaces might for example meet on a line or seam of intersection (Fig. 7.6b), and one must remember that because of the $3N-6$ vibrational modes of a general polyatomic molecule of N atoms that the interaction between electronic potential energy surfaces is multidimensional.

The observation of broadening in an absorption spectrum tells us that the photoexcited electronic state rapidly disappears, however, no further information can be

(a) (b)

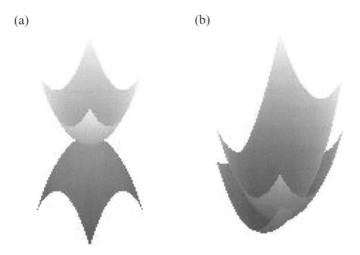

Fig. 7.6. (a) Schematic diagram of a conical intersection between two potential energy surfaces. (b) Schematic diagram of an intersection seam between two potential energy surfaces.

extracted from the absorption spectrum itself, i.e. we can neither deduce where the energy has gone nor the mechanism. But from what we have said above it should be apparent that time-resolved photoelectron spectroscopy might be a fruitful approach and, indeed, since the mid-1980s considerable efforts have been made to probe the excited state dynamics of large molecules in real time by this method. For instance, Sekreta and Reilly probed the intersystem crossing from the S_1 6^1 level of benzene to $^3B_{1u}(T_1)$ vibronic levels by a pump-probe technique with nanosecond lasers [7]. Similarly, Pallix and Colson performed one-colour [1+2] REMPI-PES on the $^1E''$ state of sym-triazine, where they found the spectra varied with the laser pulse duration [8]. Knee and co-workers extended time-resolved photoelectron spectroscopy to the picosecond time range [9]. In the same vein, Weber and co-workers examined internal conversion and intersystem crossing processes in azulene, phenanthrene, aniline, 2- and 3-aminopyridine [10], where they observed rapid changes of the photoelectron kinetic energy distributions on the picosecond time scale. Intersystem crossing in these molecules is faster than in benzene, mainly due to the existence of both the (π, π^*) and (n, π^*) states, between which a direct spin-orbit coupling is allowed. The coupling between $^1(\pi, \pi^*)$ and $^3(\pi, \pi^*)$ is much weaker. Weber *et al.* have also examined the possibility of using two ionization continua of π^{-1} and n^{-1} for probing the electronic character of an excited state, as previously suggested by Seel and Domcke [11].

The shift in the photoelectron kinetic energy upon IC and ISC can be quite large, since its magnitude is similar to the energy gap (0.1–1 eV) between the initial and final electronic states. Thus, detection of electronic dephasing from measurements of the photoelectron kinetic energy spectrum is not a difficult task, even with conventional methods. By contrast, changes in the photoelectron angular distribution (PAD) upon electronic dephasing has not been examined carefully because of the difficulty of making such measurements with conventional instrumentation. In photoelectron imaging, where the entire speed and angular distributions of the photoelectrons are observed, any variation of these two quantities can be examined quite easily.

To make this discussion more specific we shall consider the case of 1,4-diazabenzene, or pyrazine. As is well known from atomic spectroscopy, Hund's rules predict that for a two-electron orbital configuration the triplet combination will lie lower in energy than the singlet. In pyrazine the energy gap between the lowest vibrational level in the $S_1(^1B_{3u})$ state and the ground level of the lower-lying triplet state, $T_1(^3B_{3u})$ is known to be $4056\,cm^{-1}$. The vibrational density of states in the triplet configuration at this energy is such that some 20–25 vibrational levels of the triplet manifold are quasi-degenerate with the ground vibrational level of S_1. These vibrational levels are not directly excitable from the electronic ground state of pyrazine because the electric dipole selection rules for the absorption of light forbid it. However, the presence of the triplet states may be inferred by the effect of the mixing induced by vibration of the molecule close to the intersection of the singlet and triplet surfaces.

The theory of molecular radiationless transitions [12] has categorized the electronically excited state dynamics of polyatomic molecules into (a) the small molecule limit, (b) the intermediate coupling case, and (c) the statistical limit. The $S_1(^1B_{3u})$ state of pyrazine is the best-known example of the intermediate coupling case [13]. Coherent excitation of an intermediate case molecule exhibits biexponential fluorescence decay, where the fast decay is due to ultrafast ISC from an optically prepared singlet state, $|s\rangle$, into mixed singlet–triplet states and the slow decay is the depopulation of this mixed state by fluorescence to the ground electronic state. The biexponential fluorescence decay of S_1 pyrazine was extensively studied in the 1980s; however, it is noteworthy that these studies only observed the decay of the singlet state, and the dynamics in the triplet manifold remained completely dark. In principle, the triplet state can emit phosphorescence, but this is a spin-forbidden electronic transition and its yield is negligibly small. In solution it would be possible to study the transient absorption from T_1 to a higher lying triplet state, but in the gas phase the number density of triplet state molecules is too small to make direct absorption studies a feasible proposition. (N.B. in solution intermolecular relaxation processes dominate because the collision frequency with the solvent is high and this effect would mask the intramolecular electronic dephasing we are trying to study).

We have applied femtosecond time-resolved photoelectron imaging to shed light on the dark triplet manifold [14,15]. In our experiments, pyrazine, in a molecular beam, was excited to the S_1 0^0 level by a 324 nm femtosecond pump pulse, then subsequently ionized by a 197 nm femtosecond probe pulse. Ionization from the triplet state is expected to produce low-energy electrons, since triplet levels isoenergetic with the initially photoexcited singlet 0^0 level have large vibrational energies (4056 cm^{-1} in the case of T_1), and the Franck–Condon overlap, therefore, favours ionization to vibrationally excited states in the cation.

Figure 7.7 shows the photoelectron images observed at different pump-probe time delays. The cross-correlation between the pump and probe laser pulses is about 250–350 fs. The image observed at the shortest time delay consists of a number of sharp rings. However, this structure disappears with a lifetime of 110 ps. Correspondingly, a low-energy electron signal grows in the inner part of the image. The sharp rings are transitions to vibrationally excited levels in the cation, and their intensity distribution follows the Franck–Condon factor between the S_1 0^0 level

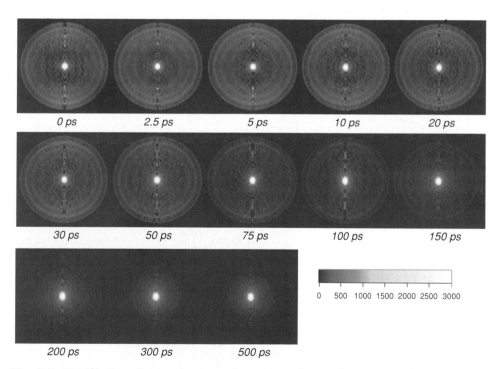

Fig. 7.7. [1+1$'$] photoelectron images of pyrazine taken at the pump-probe time delays indicated. The pump laser wavelength was fixed to the (0,0) band of S_1–S_0 transition. The images have been transformed using the Abel reconstruction method described in Chapter 3 so that they show a slice through the three-dimensional photoelectron velocity distribution. The polarization vectors of both the pump and probe lasers lie vertical in the image plane.

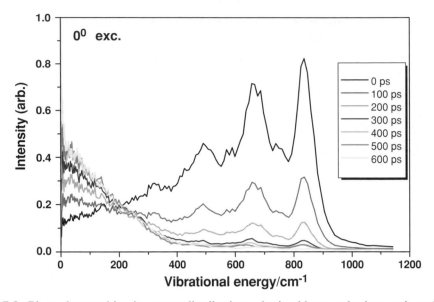

Fig. 7.8. Photoelectron kinetic energy distributions obtained by angular integration of images similar to those shown in Fig. 7.7.

and the cation. The observed lifetime of the S_1 0^0 level is in excellent agreement with the fluorescence decay lifetime of the fast component reported previously by others [9, 16]. The pump and probe polarization directions are parallel, and in the vertical direction in the figure. The photoelectron kinetic energy distributions were determined from the images, as shown in Fig. 7.8. All the spectra cross at the same energy (isosbestic point), indicating that the spectra consist of two components in dynamic equilibrium. The ionization from the triplet manifold peaks at zero kinetic energy indicating that 197 nm is not a sufficiently short wavelength to ionize the entire wavepacket in the triplet manifold, which makes the experiment relatively less sensitive to the triplet character than the singlet. We have also excited pyrazine to various vibronic levels in S_1 ($E_{\mathrm{vib}} < 2000\,\mathrm{cm}^{-1}$) and observed the photoelectron images. In all cases, the singlet signal decayed without changing its structure, which means that IVR in the S_1 manifold does not occur due to the lack of sufficient vibrational state density. The same result has been obtained for deuterated pyrazine.

In the laboratory frame, the PAD is a measure of the flux of outgoing photoelectrons as a function of angle ϑ with respect to the ionization laser symmetry axis (the electric vector in linearly-polarized light). For one-photon ionization, the definite spin and parity of the ionizing photon defines the PAD observable for a randomly oriented ensemble of molecular targets [17]:

$$\frac{d\sigma}{d\Omega} = \frac{\sigma}{4\pi}\,[1 + \beta P_2\,(\cos\vartheta\,)] \tag{7.5}$$

where, as in previous chapters, $P_K(\cos\vartheta)$ is a Legendre polynomial. The physical interpretation of the laboratory-frame PAD is particularly clear for the case of one-photon ionization of a one-electron atom. Recall the $\Delta l = \pm 1$ selection rule for an electronic transition in a hydrogen atom. We see from this that upon one-photon ionization from an s orbital, the outgoing electron will have a wavefunction of pure p character. On the other hand, ionization from a p orbital will create both s and d waves which interfere with each other. According to Bethe [18] and to Cooper and Zare [19] the anisotropy parameter is given by:

$$\beta = \frac{l(l-1)r_{l-1}^2 + (l+1)(l+2)r_{l+1}^2 - 6l(l+1)r_{l+1}r_{l-1}\cos(\eta_{l+1} - \eta_{l-1})}{(2l+1)\left[lr_{l-1}^2 + (l+1)r_{l+1}^2\right]} \quad (7.6)$$

where $r_{l\pm 1}$ are the transition dipole matrix elements for the $l \pm 1$ components of outgoing photoelectron wavefunctions, and $\eta_{l\pm 1}$ denote the phases of these waves. For an s electron, $l = 0$, the outgoing photoelectron is a pure p-wave, yielding $\beta = 2$. For other initial electron orbital angular momentum states, the magnitude of the anisotropy parameter is reduced as a result of interference between the two outgoing partial waves.

In pump-probe experiments on molecules, the ensemble of molecules excited by the pump pulse can be spatially aligned, since those ground-state molecules with the transition dipole moment (nearly) parallel to the electric vector of light are selectively photoexcited. The molecular axis alignment created by a one-photon transition with a linearly polarized pump pulse is generally expressed by:

$$F(t; \vartheta, \varphi) = A_{00}Y_{00}(\vartheta, \varphi) + A_{20}(t)Y_{20}(\vartheta, \varphi) \quad (7.7)$$

where ϑ is the polar angle of the molecular axis with respect to the pump laser polarization axis, and the $Y_{K0}(\vartheta, \varphi)$ are spherical harmonics. Notice that odd terms in K do not occur in the expansion because linearly polarized light cannot set up any orientation (preferred direction in space) of the ensemble of molecular axes (see Chapter 4). Since an ultrashort pump pulse excites the P($\Delta J = -1$), Q(0), and R(+1) branch lines coherently upon electronic transition, a nonstationary superposition of the eigenstates with different J is created. This superposition of states exhibits time-evolution of the molecular axis alignment, as expressed by the time-dependence of an alignment parameter, $A_{20}(t)$, that revives at rotational periods of a molecule. Felker and Zewail pioneered the use of the observation of these revival signals to determine the rotational level structure of large molecules that are difficult to study in the frequency domain [20].

In a [1+1'] pump and probe ionization experiment with laser polarization vectors parallel to each other, the PAD is characterized by:

$$\frac{d\sigma}{d\Omega} = \frac{\sigma}{4\pi}\left[1 + \beta_2(t)P_2(\cos\vartheta) + \beta_4(t)P_4(\cos\vartheta)\right] \quad (7.8)$$

where the anisotropy coefficients are related to the time-dependent alignment parameters by

$$\beta_L(t) = \sum_K P_{KL} A_{K0}(t).\qquad(7.9)$$

The functions P_{KL} involve cross-terms of the dipole transition matrix elements inducing interference, as in the case of a one-electron atom. Notice that as the alignment parameter $A_{20}(t)$ varies with the rotational wave packet motion, the anisotropy parameters β varies as well. Thus, the PAD is sensitive to the rotational dynamics that change the alignment or orientation of the molecule as well as to processes such as IC and ISC that change the electronic character of the molecule.

In the case of pyrazine, the photoelectron speed and angular distribution can be divided into the singlet and triplet parts,

$$\frac{d^2\sigma}{dvd\Omega} = \frac{\sigma_s(v)}{4\pi}\big[1 + \beta_2^s(t)P_2(\cos\vartheta) + \beta_4^s(t)P_4(\cos\vartheta)\big]$$
$$+ \frac{\sigma_T(v)}{4\pi}\big[1 + \beta_2^T(t)P_2(\cos\vartheta) + \beta_4^T(t)P_4(\cos\vartheta)\big].\qquad(7.10)$$

Figure 7.9 shows polar plots of the PAD recorded by $[1+1']$ ionization of pyrazine measured with the probe light aligned parallel to the pump laser polarization. The distributions are for the singlet components. The PAD at $t = 0$ shows a characteristic four-fold distribution that clearly indicates that the photoelectrons are ejected not only along the probe laser polarization but also perpendicular to it. The transition dipole moment of the S_1–S_0 transition of pyrazine is perpendicular to the molecular plane. Thus the x-axes of the photoexcited molecules are predominantly aligned parallel to the laser polarization at $t = 0$, so that the molecular plane is perpendicular to the laser polarization. The PAD at $t = 0$ shows that although the electrons are ejected predominantly out of the plane some are ejected in the plane of the molecule.

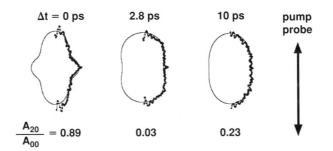

Fig. 7.9. Photoelectron angular distributions obtained by radial integration of the Abel-transformed images recorded by $[1+1']$ photoionization for various time delays. The integration was performed for a particular radial part, corresponding to the photoelectron kinetic energy of 600–900 meV, to extract the PAD in ionization from the singlet state.

At $t = 2.8$ ps, after the pump pulse, the alignment parameter for pyrazine at a rotational temperature of 20 K almost vanishes, so the ensemble of excited state molecules at this time becomes randomly oriented in space, and the PAD loses the four-lobe structure exhibited at $t = 0$. The PAD does not become isotropic however because although the axes of the excited state molecules are randomly aligned the photoionization cross-section depends on the alignment of the molecules with respect to the probe laser polarization, and consequently the PAD exhibits an angular dependence.

The results just described demonstrate that pump-probe photoelectron spectroscopy allows us to observe the PAD for a molecular ensemble that is momentarily aligned in space, and it provides some insight into the photoionization dynamics of molecular orbitals fixed in space. Such a measurement has been regarded for some time as a highly desirable experimental goal in photoionization studies of molecules. It has been achieved previously by coincidence detection of daughter ions and electrons in dissociative ionization, but this approach cannot be generalized to large molecules. An alternative approach is to measure the PAD for an ensemble of aligned or oriented molecules. In this vein, McKoy [21], Zare [22] and coworkers studied in detail the photoionization of laser-aligned NO molecules in the $A\,^2\Sigma$ state. The creation of a rotational wavepacket by a short pulse and subsequent measurement of the PAD opens a new way to achieve measurements of molecular frame PADs.

7.3.2 Time-resolved photoelectron photoion coincidence imaging

When two independent detectors are used, one can identify, in principle, both the positive ion and the negative electron that are created simultaneously upon ionization. This so-called coincidence technique has been usefully applied, for instance, to study the excited state dynamics of neutral clusters. In the case of cluster ions, one can mass-select a particular size of cluster prior to a pump-probe experiment. However, it is difficult to mass-select neutral clusters. Spectroscopic selection of the cluster size is sometimes possible by tuning the high-resolution laser wavelength, since the vibrational and electronic spectra of clusters are size dependent. However, a pump-probe experiment using femtosecond lasers with, necessarily, a large band-width does not have this option.

Radloff and co-workers have used a photoelectron photoion coincidence (PEPICO) technique in conjunction with femtosecond pump-probe methods to study excited state dynamics of neutral clusters. In a study of benzene clusters [23], they have shown that for both the monomer and the dimer, the S_2 state relaxes within 40 fs by internal conversion to S_1 and S_0, and that the conversion to the S_0 greatly exceeds that to S_1. The ratio of these processes was $100:1$ in the monomer and 40 : 1 in the dimer. The S_1 state, populated by internal conversion from S_2, further decays down to S_0 but on a slower time scale; 6.7 ps in the monomer and >100 ps in the dimer.

Hayden and coworkers [24] have further extended femtosecond pump-probe co-incidence methods by introducing delay line detectors that allow detection of the arrival time and position of both an ion and an electron. This rather complicated experimental method is able to fully correlate the mass and velocity of a pho-toion and the velocity of the concomitantly ejected electron. The method has been demonstrated by observing the photodissociation of NO_2 followed by photoion-ization of the NO fragment. In these experiments the photoelectron recoil velocity was correlated with the recoil velocity of NO^+. At short pump-probe delays, the electron apparently disfavours ejection towards the counterpart O atom side, giving rise to an asymmetric MF-PAD, whereas ionization of the NO fragment after the N–O bond rupture, i.e. at long time delays, showed a symmetric PAD, as expected for an isolated NO molecule. This technique potentially provides highly detailed in-formation on the photodissociation dynamics, but its full exploitation awaits further experimental efforts.

In coincidence experiments, the mean number of electron–ion pairs created per laser pulse has to be very small, $\ll 1$, in order to suppress the rate of false coinci-dences. In the experiment by Radloff and coworkers, the total rate of coincidence was 0.05 per pulse with 10% of this fraction being false coincidences. The low count rate is a particular difficulty for coincidence experiments, and it is only by employing high repetition rate (> kHz) lasers that these experiments are possible at all (although one should note that this type of experiment does not require a high photon flux, so the energy per pulse may be $\sim$100 nJ or less).

7.4 Time-resolved photoion imaging

Excitation of the molecule to an unbound potential energy surface creates a wavepacket, which spreads over a large phase space volume, and the system tends to exhibit statistical behaviour. Experimental investigations of these sorts of reac-tions are mainly based on the measurements of reaction rates and comparison with statistical theories such as Rice–Ramsperger–Kassel–Marcus (RRKM) theory. A particularly important class of unimolecular reactions, from the point of view of combustion science, concerns the break-up of radical species. In order to investi-gate the dynamics of these systems, we need an experimental method that allows efficient formation of radicals in a range of highly-excited states and measures the subsequent decay for each internal energy. An experimental method that comprises these two features is time-resolved photoion imaging [14]. The principle is illus-trated in Fig. 7.10. The radicals R are produced by photodissociation of a parent

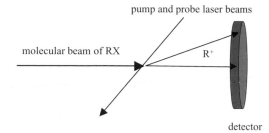

Fig. 7.10. Photoion imaging of unimolecular decay of radicals. Radicals, R, are produced by photodissociation of molecules RX in a molecular beam following irradiation by the pump laser pulse. The radicals are ejected from the molecular beam with various internal energies. The recoil energy of the radical away from the primary beam is related to the internal energy of R through the energy and momentum conservation laws. Radicals with small internal energies have large translational energy and are deflected most. The radicals are imaged by ionizing them with a probe pulse and detecting the ions on a two-dimensional detector.

molecule RX in a molecular beam. Following irradiation by the pump laser pulse radicals are ejected from the molecular beam with various internal energies. With a certain pump-probe delay, these radicals are interrogated by multiphoton ionization using a probe laser pulse. When the atom, X, is created in a single quantum state, the translational energy of the radical, R, is related to the internal energy of R through the energy and momentum conservation laws. Then, radicals with small internal energies have large translational energy and are deflected far from the molecular beam, while radicals with large internal energy have smaller translational energy and remain close to the molecular beam. In this way, radicals with different internal energies are dispersed in space and can be detected by a two-dimensional position-sensitive detector. If the radical has a sufficient internal energy to surmount the barrier for secondary dissociation, $R \rightarrow A + B$, it will decompose before being ionized by the probe laser pulse. Two-dimensional ion imaging can follow the disappearance of radicals with a certain translational energy. A series of ion images, taken by varying the pump-probe delay, allows observation of the unimolecular decay of energy-dispersed radicals. The novel feature of the ion imaging method for investigating these reactions is the dispersion of the radical R with different internal energies, which is the key with which to unlock the connection between the decay rate and the internal energy, E.

Shibata *et al.* have applied this method to study the unimolecular dynamics of CH_3CO produced by 255 nm photodissociation of CH_3COCl [14]. Ion images of CH_3CO as a function of pump-probe time delay were measured. From these the translational energy distribution of acetyl radicals was obtained; that is the internal energy distribution of 'surviving' acetyl radicals. Unsurprisingly, radicals with higher internal energies were found to decay faster. More quantitatively, the decay rates at each internal energy could be estimated by least-squares fitting of single or double exponential decays to each plot. The observed rate was nearly one order of magnitude smaller than the RRKM calculations.

The discrepancy is thought to be due to restricted intramolecular vibrational relaxation (IVR) and rotational metastability. If the initial internal energy is partitioned into vibrational modes that are poorly coupled to the C–C stretching modes, IVR becomes incomplete, leading to a smaller reaction rate than predicted by RRKM theory. Since acetyl chloride becomes pyramidal in the $^1(n, \pi^*)$ state, C–Cl repulsion is expected to excite molecular rotation of CH_3CO around the a-axis. The a-axis rotation is poorly coupled with the reaction coordinate, and if K-scrambling is restricted, it will make CH_3CO rotationally metastable.

Recently, Martinez-Nunez and Vazquez [25] investigated this process for the same energy range by classical trajectory calculations. They found intrinsic RRKM behaviour for the dissociation with zero total angular momentum. However, when molecular rotation was excited around any of the three principal axes the dissociation rate decreased irrespective of the direction of the rotational axis. In addition, torsional excitation of the methyl group diminished the dissociation rate significantly. At $96\,kJ\,mol^{-1}$ and for 6.7 ps, which is the average lifetime at this energy, K-scrambling was found to be 15–40%, suggesting that the rotation–vibration coupling is fairly weak. Thus, they concluded that torsional excitation of CH_3 and a-axis rotation is the cause of the observed non-RRKM behaviour, in accord with the earlier speculation.

7.5 Conclusions

This chapter has reviewed some of the recent progress that has been made in pump-probe imaging on the sub-picosecond time scale. Time-resolved photoelectron imaging enables us to observe electronic, vibrational and rotational dynamics of molecules. The method is particularly useful in probing the entire reaction dynamics from its Franck–Condon region to the products all the way along the reaction coordinate including nonadiabatic transition in the process. The photoion imaging experiment provides data on the time-dependent photofragment kinetic energy distributions. This can occur both in the formation and dissociation of radicals. If the parent molecule ABC dissociates into AB+C in competition with IVR in ABC, the translational energy distribution of AB will vary, most likely towards a lower energy as a function of time. If AB dissociates into A+B, as in the case of CH_3CO, the fragments with low translational energy will diminish as a function of time.

We have not been able cover every application of imaging with femtosecond lasers. An important area we have neglected concerns the fact that even for relatively modest pulse energies the peak power present in a femtosecond pulse can be extremely large, of order $TW\,cm^{-2}$ or higher. Thus the instantaneous electric field felt by the molecule may approach, or even exceed, that which binds the valence electrons to the nuclei, particularly if the beam is focused. This can lead to some extremely interesting, but sometimes unwanted, effects. In a strong electric field the energy levels of a molecule are shifted and split by the Stark effect, and in extreme cases the electric field can even cause molecular ionization. Imaging techniques have been developed to study these processes [26], but more importantly in the context of this chapter is that the experimentalist wishing to apply femtosecond lasers to imaging needs to be aware of their existence.

Acknowledgements

The authors are grateful to the Ministry of Education, Culture, Sports, Science and Technology of Japan (contract 14204063) and to the European Union (IMAGINE contract ERB 4061 PL 97-0264) for financial support.

References

1. D. E. Spence, P. N. Keam, W. Sibbett, *Opt. Lett.* **16**, 42 (1991).
2. P. W. Atkins, R. S. Friedman, *Molecular Quantum Mechanics*, 3 edn. (Oxford University Press, Oxford, 1997).
3. K. L. Sala, G. A. Kenney-Wallace, G. E. Hall, *IEEE J. Quant. Elect.*, **16**, 990 (1980).
4. A. Stingl, C. Spielmann, F. Krausz, R. Szipocs, *Opt. Lett.* **19**, 204 (1994).
5. P. Kruit, F. H. Read, *J. Phys. E - Sci. Instrum.* **16**, 313 (1983).
6. D. W. Chandler, D. H. Parker (eds.) *Velocity Mapping of Multiphoton Excited Molecules*, (Wiley, New York, 1999), vol. 25.
7. E. Sekreta, J. P. Reilly, *Chem. Phys. Lett.* **149**, 482 (1988).
8. J. B. Pallix, S. D. Colson, *Chem. Phys. Lett.* **119**, 38 (1985).
9. J. L. Knee, F. E. Doany, A. H. Zewail, *J. Chem. Phys.* **82**, 1042 (1985).
10. P. M. Weber, N. Thantu, *Chem. Phys. Lett.* **197**, 556 (1992); B. Kim, C. P. Schick, P. M. Weber, *J. Chem. Phys.* **103**, 6903 (1995).
11. M. Seel, W. Domcke, *Chem. Phys.* **151**, 59 (1991).
12. M. Bixon, J. Jortner, *J. Chem. Phys.* **48**, 715 (1968).
13. P. Avouris, W. M. Gelbert, M. A. El-Sayed, *Chem. Rev.* **77**, 793 (1977).
14. T. Shibata, T. Suzuki, *Chem. Phys. Lett.* **262**, 115 (1996).
15. T. Suzuki, L. Wang, H. Kohguchi, *J. Chem. Phys.* **111**, 4859 (1999); T. Suzuki, *Abstr. Papers Am. Chem. Soc.* **218**, 145 (1999); L. Wang, H. Kohguchi, T. Suzuki, *Faraday Discussions* **113**, 37 (1999); M. Tsubouchi, B. J. Whitaker, L. Wang, H. Kohguchi, T. Suzuki, *Phys. Rev. Lett.* **86**, 4500 (2001); J. K. Song, M. Tsubouchi, T. Suzuki, *J. Chem. Phys.* **115**, 8818 (2001).
16. D. B. McDonald, G. R. Fleming, S. A. Rice, *Chem. Phys.*, **60**, 335 (1981); I. Yamazaki, T. Murao, T. Yamanaka, K. Yoshihara, *Faraday Discussions* **75**, 395 (1983); A. Lorincz, D. D. Smith, F. Novak, R. Kosloff, D. J. Tannor, S. A. Rice, *J. Chem. Phys.* **82**, 1067 (1985).
17. C. N. Yang, *Phys. Rev.* **74**, 764 (1948).
18. H. A. Bethe. *Handbuch der Physik* (Springer-Verlag, Berlin, 1933), vol. 24, pp. 483–4.
19. J. Cooper, R. N. Zare, *J. Chem. Phys.* **48**, 942 (1968).

20. P. M. Felker, A. H. Zewail, *J. Chem. Phys.* **86**, 2460 (1987); J. S. Baskin, P. M. Felker, A. H. Zewail, *J. Chem. Phys.* **86**, 2483 (1987).
21. K. S. Wang, V. McKoy, *Annu. Rev. Phys. Chem.* **46**, 275 (1995).
22. D. J. Leahy, K. L. Reid, R. N. Zare, *J. Phys. Chem.* **95**, 8154 (1991); D. J. Leahy, K. L. Reid, R. N. Zare, *J. Chem. Phys.* **95**, 1757 (1991); D. J. Leahy, K. L. Reid, H. K. Park, R. N. Zare, *J. Chem. Phys.* **97**, 4948 (1992); K. L. Reid, D. J. Leahy, R. N. Zare, *J. Chem. Phys.* **95**, 1746 (1991); K. L. Reid, D. J. Leahy, R. N. Zare, *Phys. Rev. Lett.* **68**, 3527 (1992).
23. W. Radloff, V. Stert, T. Freudenberg, I. V. Hertel, C. Jouvet, C. Dedonder-Lardeux, D. Solgadi, *Chem. Phys. Lett.* **281**, 20 (1997).
24. J. A. Davies, J. E. LeClaire, R. E. Continetti, C. C. Hayden, *J. Chem. Phys.* **111**, 1 (1999).
25. E. Martinez-Nunez, S. A. Vazquez, *Chem. Phys. Lett.* **316**, 471 (2000).
26. T. Suzuki, B. J. Whitaker, *Int. Rev. Phys. Chem.* **20**, 313 (2001).

Part Two

Applications

8

Kinematically complete imaging of molecular many-body fragmentation: coincident multi-particle detection and analysis

ULRICH MÜLLER AND HANSPETER HELM

8.1 Introduction

The break-up of a molecule into neutral products plays an important role in many astrophysical, atmospheric and plasma processes. Knowledge of reaction rates are frequently sufficient for modelling complex environments such as molecule formation in the interstellar medium. However a strong discrepancy between theory and experient for a specific reaction rate or else the curiosity of the prudent researcher often calls for more detail on the molecular dynamics which is hidden in a rate coefficient. Such a case is discussed in this contribution. We examine neutral tri-atomic hydrogen and show the great potential of modern imaging technologies to elucidate molecular dynamics at an unprecedented level of detail. The H_3 molecule is the reaction intermediate in dissociative recombination (DR) of slow electrons with H_3^+.

$$H_3^+ + e \rightarrow H_3 \rightarrow H_2(v, J) + H(1s)$$
$$\rightarrow H(1s) + H(1s) + H(1s)(1). \qquad (8.1)$$

Remarkable discrepancies of many orders of magnitude exist among different experiments and between theory and experiment for the rate of DR [1]. We highlight below the power of advanced imaging technology to uncover details of molecular dissociation active in reaction 8.1.

Observing neutral fragments which slowly recede from the centre-of-mass, as a consequence of photoexitation of an unstable molecular state, permits the study of the correlation of the fragment momentum vectors. This correlation reflects the decay dynamics on the repulsive potential energy surface as well as the nuclear and electronic geometry in which the molecular system enters the dissociative continuum. For large molecules, dissociation into more than two massive fragments may constitute a major decay pathway [2–6] but this is largely uncharted territory of chemistry and physics at the time of writing.

189

In order to pinpoint details of molecular dynamics in such processes, the parent molecule has to be prepared in a well-characterized initial quantum state. This preparation is readily achieved by laser excitation of the system to a predissociative excited state. Provided the lifetime of the excited state is sufficiently long, the excitation spectrum separates into isolated lines, each characterized by a specific set of quantum numbers. This isolated set of coordinates of the laser-prepared initial state can be mapped into a well-characterized set of coordinates of final products using the imaging tool described here. The map obtained provides us with the most detailed view of molecular dynamics quantum mechanics allows us to obtain.

The study of photodissociation dynamics by laser preparation of a well-defined initial quantum state has a long history in translational spectroscopy [7–9] as well as in momentum spectroscopy using position-sensitive detectors [10,11]. Positive and negative molecular ions [12–15] as well as neutral molecules [16–19] have been investigated using these techniques. Molecules larger than diatomics have only rarely been the subject of such studies [2,20–22] due to the difficulty in monitoring more than two massive fragments.

Extensive efforts undertaken at the University of Freiburg over the past 5 years have led to a versatile multi-particle coincident imaging tool which achieves a kinematically complete description of the final state. This multi-hit coincidence experiment is based on time- and position-sensitive particle detectors which combine large detection areas with high spatial and temporal resolution. Such detectors, originally pioneered in work of the FOM-group [23], have seen remarkable development over the past 15 years [24–28]. Their implementation in the Freiburg experiment combines detection electronics and on-line analysis algorithms which represent a substantial improvement over detectors which have recently been employed in atomic double- and multiple-ionization studies [29,30] and their predecessors which found application in Coulomb explosion experiments [31–35].

In the following, we first describe the fast-beam translational spectrometer recently developed in Freiburg [21]. Then, we describe the multi-hit readout, and the data reduction algorithm [22] to determine the fragment momentum vectors in the centre-of-mass frame (c.m.). We discuss suitable projections of the multi-dimensional (6-D) photodissociation cross-section to gain insight into the decay dynamics. The performance of the apparatus and the data evaluation procedure is illustrated at the example of photodissociation of the triatomic hydrogen molecule H_3. We present a kinematically complete final state analysis of the two-body decay into $H + H_2$ as well as the three-body decay into three hydrogen atoms $H(1s) + H(1s) + H(1s)$ and address its significance for the DR-process in reaction 8.1.

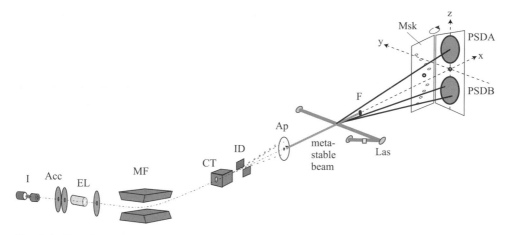

Fig. 8.1. Fast-beam translational spectrometer for the investigation of molecular many-body dissociation processes.

8.2 Experimental

8.2.1 Fast beam and laser excitation

Figure 8.1 shows the Freiburg fast-beam photofragment spectrometer for the detection of molecular many-body fragmentation. Molecular ions are created in an ion source (I), accelerated to a translational energy of several keV (Acc), mass selected (MF), and focused by an Einzel lens (EL). A small fraction of the ions is neutralized by charge transfer. After the charge-transfer cell (CT), the unreacted ions are removed by an electric field (ID), and the products of dissociative charge transfer are stopped by the aperture (Ap). A fast, well-collimated beam of neutral molecules is also created in the charge transfer process, propagating along the x-direction in the laboratory frame. The dissociation into n fragments is initiated in a primary process such as laser-excitation or atomic collision. The undissociated part of the parent neutral molecular beam is intercepted by a miniature beam flag (F). The fragments separate from each other due to their kinetic energy in the c.m. frame. They are recorded in coincidence by a time- and position-sensitive multi-hit detector (PSDA, PSDB) after a free-flight of about 150 cm. The positions and the arrival time differences of the fragments in the plane of the detector are individually measured.

In laser-excitation experiments, a home-built standing wave dye laser (Las) pumped by an argon-ion laser is used. To take advantage of its high intracavity power, the neutral beam is crossed with the laser beam inside the cavity. The vacuum chamber is equipped with Brewster windows to minimize the cavity losses. The laser wavelength is controlled by a birefringent filter which is tuned by a stepping motor. For operation at a fixed wavelength, an intracavity etalon is inserted to reduce the laser bandwidth to about $0.1\,\mathrm{cm}^{-1}$. The laser beam can be turned on and off by an intracavity shutter operated under computer control. This allows us to collect the events from metastable decay of the molecules separately from laser-induced fragmentation.

8.2.2 Detector requirements and specifications

A sophisticated time- and position-sensitive detection system is required to measure the products of a multi-particle dissociation process in coincidence. The sensitive detector area should be comparatively large ($\approx 10\,\mathrm{cm}$ outside diameter) to achieve a high geometric collection efficiency. To cover a significant part of the phase space of fragmentation configurations, we can discriminate consecutive fragment impacts with temporal and spatial separations as low as 20 ns and 1 mm, respectively. The impact coordinates in the detector plane and the arrival time differences are determined with accuracies of better than $100\,\mu\mathrm{m}$ and 100 ps, respectively.

Common to all detector concepts in use are micro channel plates (MCP) which multiply secondary electrons produced by the impact of fast neutrals, photons or charged particles. They achieve amplifications of $10^6 \ldots 10^8$, retaining the position and time information of the original impact with accuracies of $25\,\mu\mathrm{m}$ and 100 ps, respectively. Several methods can be considered to read out the position and time information. Position-sensitive anodes based on charge division (resistive anodes [36,37], capacitor chains [23,24,38], wedge-and-strip anodes [39]) achieve good spatial resolution but have comparatively long recovery times (several microseconds). Such anodes are not suitable for multi-hit applications, unless the sensitive area is subdivided and covered by several independent detector systems which greatly increases the hardware expenses. Using multi-element anodes with individual readout circuits for each element, excellent multi-hit capabilities have been demonstrated [26] but the spatial resolution was only about 1 mm. An imaging detector which combines a phosphor screen and a CCD-camera with multi-anode photomultipliers has been developed by Amitay and Zajfman [27]. This detector offers multi-hit capabilities and good time and position resolution, but the total event rate is limited by the relatively low camera readout rate (50 frames/s).

8.2.3 Multi-hit methods and readout system

The time- and position-sensitive multi-hit detector developed in Freiburg consists of two units. Figure 8.2 shows one of them schematically. Two Z-stacks of MCPs with 50 mm outside diameter (44 mm diameter active area) are mounted in the y–z plane of Fig. 8.1, centred at distances of 35 mm from the x-axis. The electron charge cloud from the MCPs is collected by a position-sensitive delayline anode (PSA) separated by 15 mm from the back surface of the MCP. The shaping element (SE) is biased

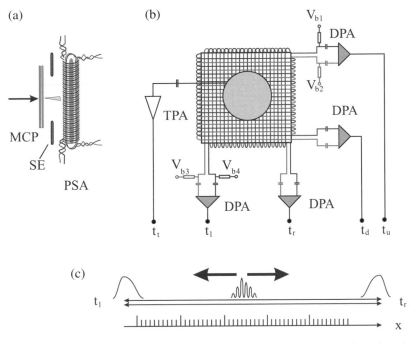

Fig. 8.2. Multi-hit time- and position-sensitive detector unit. Schematic side view (a), front view (b). The principle of the position measurement is shown in (c). MCP, multichannel plates; SE, shaping electrode; PSA, position sensitive delayline anode; TPA, fast preamplifier for timing signal; DPA, differential preamplifiers for position signals.

appropriately to minimize distortions of the extraction field at the boundary of the MCP. Position sensitive delayline anodes were originally introduced by Sobottka and Williams [25]. The delayline anodes in our apparatus have been developed by Jagutzki *et al.* [28]. As shown in Fig. 8.2, two waveguides are formed by pairs of parallel copper wires. They are wrapped in two layers around a conductive base plate, held in place and separated from each other and the base plate by ceramic spacers. A bias voltage between the wire pairs leads to an imbalance of the amount of charge collected by the wires. Electromagnetic pulses due to the charge impact propagate to the terminals of the waveguides which are capacitively coupled to differential preamplifiers (DPA). The delayline anode transforms the position coordinate of the electron cloud into arrival time coordinates of electromagnetic pulses. The principle of the position measurement is visualized in Fig. 8.2c for one dimension.

The computer-controlled readout system for the time- and position-sensitive detectors is given schematically in Fig. 8.3. The output pulses of the differential and the timing amplifiers (PA) are converted to NIM-standard pulses by constant-fraction (CF) discriminators. To measure the pulse arrival times, we use three

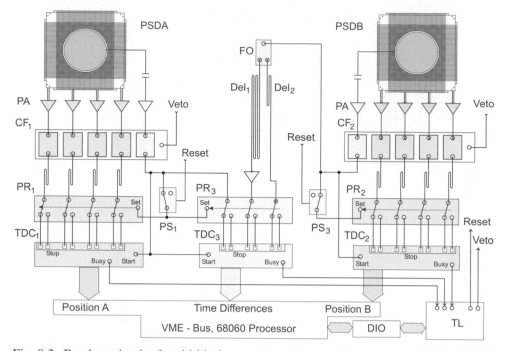

Fig. 8.3. Readout circuit of multi-hit time- and position-sensitive detector. PSDA, PSDB, time- and position-sensitive delayline detector units; PA, preamplifiers; CF, constant-fraction discriminators; FO, pulse amplifier (fan-out); Del, cable delays; PR, pulse routers; PS, pulse selectors; TDC, time-to-digital converters; TL, trigger logic; DIO, digital input/output interface.

time-to-digital converters (TDC) with eight channels each. TDC_1 and TDC_2 measure the position information of detectors PSDA and PSDB, respectively. TDC_3 measures the time differences between events. The time resolution can be hardware selected in the range between 25 and 125 ps. For the detection of multiple hits on both detector units, pulse routers (PR in Fig. 8.3) are provided for all position and timing signals. A description of the pulse routers (PR) developed in Freiburg can be found in Ref. [21]. The conversion activity of the TDCs is synchronized by a trigger logic (TL). The control of the experiment, the data readout, and the pre-analysis are performed by a dedicated processor (68060).

To calibrate the position scale, a stainless steel mask (Msk) with a precisely machined hole pattern is placed in front of the detector. The mask is mounted on a hinge and can be placed and retracted by a rotary feedthrough under vacuum. The timing TDC is calibrated with a precision time interval counter. This calibration information is stored and allows us to transform online the binary raw data of the TDCs into the hit positions and arrival time differences of the fragments in the laboratory frame.

8.2.4 Data reduction algorithms

Data reduction algorithms are required to extract the physically relevant information from the measured quantities in a fast beam experiment. For two-body decay, approximate formulas were developed by de Bruin and Los [23] to calculate the kinetic energy release W and the angle of fragment ejection. General data reduction strategies for multi-particle fragmentation into an arbitrary number n of fragments have been developed by Beckert and Müller [22]. The $3n$ cartesian components of the fragment velocity vectors in the laboratory frame are determined by an iterative algorithm from the $2n$ spatial impact coordinates, the $n - 1$ arrival time differences, the known translational velocity of the parent beam, and the projection of the flight length of fragments on the axis (assumed to be identical to the distance from the interaction region to the detector centre). The redundant pieces of information are used to assign the fragment masses and to determine the $3n - 3$ independent components of the momentum vectors in the c.m. frame. With only three iteration steps, the numerical accuracy of the new algorithm [22] is far better than the experimental uncertainty of the raw data. Three-body decay with equal fragment masses and two-body decay are special cases of this general data reduction procedure. In these special cases, the transverse components of the centre-of-mass velocity can additionally be determined from the redundant information and used as a consistency check to suppress false coincidences. The new algorithm is also superior to the frequently used approximate formula [23] for two-body decay.

8.3 The H_3 molecule

In order to illustrate the molecular dynamics study undertaken with this imaging tool, we first need to address specific features of neutral triatomic hydrogen. In Fig. 8.4, we show the energy levels of H_3 relevant for the investigations in this chapter. The electronically excited states of H_3 (except for the repulsive 2p $^2E'$ ground state surface) can be viewed as Rydberg states with a tightly bound H_3^+ core [40]. The equilibrium geometry is that of an equilateral triangle. The states of H_3 are labelled [41–45] by $nL\lambda\Gamma\{v_1, v_2\}^{l_2}(N, G)$. The electronic states are characterized by the principal quantum number n, the electronic angular momentum L, the projection of L onto the molecular top axis λ, and the electronic symmetry Γ in the D_{3h} group. The nuclear vibrational motion is labelled by the quantum numbers of the symmetric stretch (v_1) and degenerate (v_2) modes, and the vibrational angular momentum l_2. The molecular rotation is described by the total angular momentum N disregarding electron spin, the projection on the molecular top axis K, and Hougen's convenient quantum number [46] $G = l_2 + \lambda - K$. The lowest rotational level $N = K = 0$ of the 2p $^2A_2''$ electronic state is the only metastable state [43] of H_3. Molecules in

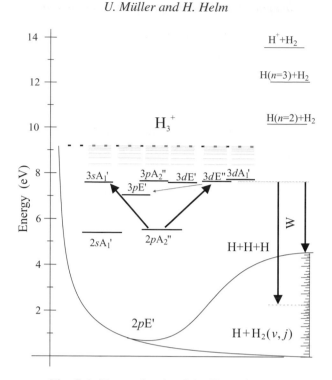

Fig. 8.4. Energy levels of the H_3 molecule.

this state can be prepared with high efficiency by near-resonant charge transfer of rotationally cold H_3^+ in Cs. Starting from this metastable state, we can populate individual excited states by laser-photoexcitation. Here, we restrict the discussion to the 3s $^2A_1'\{v_1 = 0, v_2 = 0\}$ ($N = 1, G = 0$) and 3d $^2E''\{v_1 = 0, v_2 = 0\}$ ($N = 1, G = 0$) Rydberg states of H_3 which are reached at frequencies of 16695 and 17297 cm^{-1}, respectively. Corresponding states with one quantum of vibrational excitation in the symmetric stretch or the degenerate mode can also be excited. We will now present measurements of the vibrational ground states only. For simplicity, we sometimes omit the rotational and vibrational labelling of the initial states.

The laser-prepared states couple to the continuum of the repulsive ground state potential energy surface and dissociate within 3 to 10 ns [18] by two-body decay into H + H_2 or by three-body decay into three H(1s) hydrogen atoms. In the two-body decay channel, the kinetic energy W of the fragments is the difference between the energy of the initial state and the rovibrational energy of the $H_2(v, J)$ fragment and it assumes discrete values. In the three-body decay, the sum of the kinetic energy of the three fragments is fixed at the energy difference between the excited state and the H(1s) + H(1s) + H(1s) limit. The sharing of the energy among the fragments is controlled by the decay dynamics. A radiative transition to the

Franck–Condon region of the ground state surface with subsequent decay into $H + H_2$ is also energetically possible. Naturally, this process produces a continuum of fragment kinetic energies. After this detour into the specific molecular background to our example, we return to the topic of imaging.

8.4 Imaging two-body decay processes

Coincidences between fragment pairs on the upper (PSDA) and the lower (PSDB) detector areas can be produced either by two-body decay or by three-body decay of H_3. For each detected event, we determine the quantity

$$Z = R_1/R_2$$

where R_1, R_2 are the distances of the fragment hits from the detector centre. The spectrum of Z following photodissociation of the H_3 $3s^2A_1'$ state is shown in Fig. 8.5a. In two-body decay, the value of Z very closely approximates the mass ratio of fragments. For H_3, the events which fall into the peaks at $Z = 0.5$ or $Z = 2.0$ are accepted as two-body decay. We determine the kinetic energy release W in the c.m. frame for these events. A spectrum of W from two-body decay of the

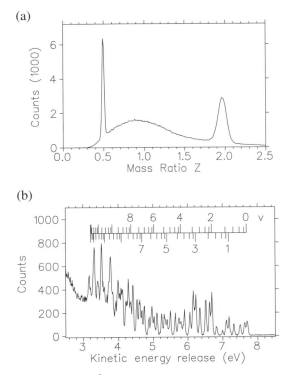

Fig. 8.5. Two-body decay of H_3 $3s\,^2A_1'$ $\{v_1 = 0, v_2 = 0\}$ ($N = 1$, $G = 0$). (a) Spectrum of the mass ratio Z, (b) spectrum of the kinetic energy release W.

$3s\,^2A_1'$ state is shown in Fig. 8.5b. Pronounced discrete peaks appear in the
W-spectrum. They correspond to the rovibrational states of the $H_2(v,J)$ fragment as
indicated by the stick spectrum in Fig. 8.5b. The current energy resolution in our
apparatus is 50 meV (FWHM) at 5 eV which is appreciably better than the results
achieved in previous investigations [19]. The spectrum can be deconvoluted using
basis functions determined by a Monte Carlo simulation which takes into account
the finite geometric collection efficiency of our photofragment detector. The final
state rovibrational population of the $H_2(v,J)$ fragment is shown by the size of the
squares in Fig. 8.6a. The corresponding distribution of the $3d\,^2E''$ initial state is
shown in Fig. 8.6b.

(a)

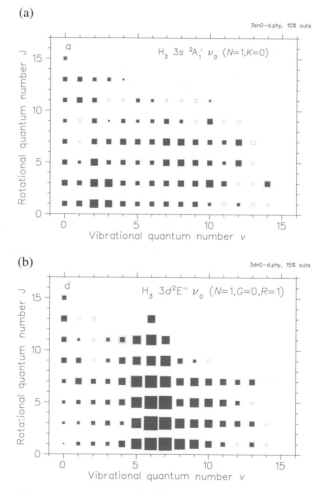

(b)

Fig. 8.6. Rovibrational distributions of the $H_2(v,J)$ fragment produced by two-body decay
of the H_3 molecule in the (a) $3s\,^2A_1'\{v_1 = 0,\, v_2 = 0\}(N = 1,\, G = 0)$ and (b) $3d\,^2E''\{v_1 = 0,\, v_2 = 0\}(N = 1,\, G = 0)$ states.

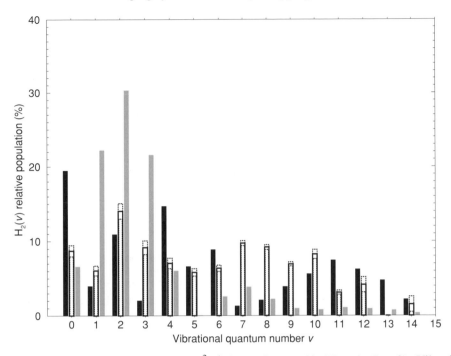

Fig. 8.7. Two-body decay of H_3 $3s$ $^2A_1'$ $\{v_1 = 0, v_2 = 0\}$ ($N = 1, G = 0$): Vibrational population of the H_2 fragment. Legend: black bars, theory by Schneider and Orel [48]; grey bars, theory by Orel and Kulander [47]; open bars, this experiment.

The highly structured distributions have no similarity with thermal rotational or vibrational distributions. Note that these distributions each arise from dissociation of a well-characterized single rovibrational initial state. We find that the $H_2(v,J)$ final state populations depend sensitively on the electronic initial state and reflect the mechanism of predissociation as well as the subsequent decay dynamics on the repulsive potential energy surface. The fragmentation process can only be understood by a quantum-dynamics calculation.

In Fig. 8.7, we compare the measured vibrational distributions [19] (open bars) to the results of two-dimensional wave packet calculation by Orel and Kulander [47] (grey bars) and by Schneider and Orel [48] (black bars). The new calculations [48] are in remarkably good agreement with the experiment considering that the rotational degree of freedom cannot yet be tackled by theory. To achieve this high quality, the calculations had to explicitly include the non-adiabatic couplings between the potential energy surface of the laser-prepared initial state and the repulsive ground state. Only recently have such comprehensive calculations become possible due to the progress in computer power. Agreement between experiment and theory has also been achieved for the $2s$ $^2A_1'$ and the vibrationally excited $3s$ $^2A_1'$ states of H_3.

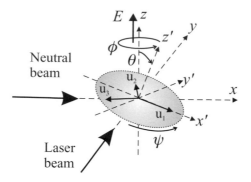

Fig. 8.8. Parameterization of the fragment momentum vectors in a three-body decay process.

8.5 Imaging three-body decay processes

The broad continuum in the spectrum of the distance ratio Z in Fig. 8.5a is due to three-body decay where one of the fragments remains unobserved. As shown by Müller and Cosby [20], a certain amount of insight into the three-body decay of H_3 can be gained by measuring two of the three fragments in coincidence and by modelling and evaluating the apparent kinetic energy release spectra. However, a direct analysis is possible with the new Freiburg photofragment spectrometer, in which triple coincidences are collected, and the momenta of the three hydrogen atoms $m\mathbf{u}_1$, $m\mathbf{u}_2$, and $m\mathbf{u}_3$ in the center-of-mass frame (see Fig. 8.8) are determined individually. Momentum conservation requires that $\mathbf{u}_1 + \mathbf{u}_2 + \mathbf{u}_3 = 0$, which implies that the three momentum vectors have only six independent components. With this spectrometer, we determine the six-fold differential cross-section, completely characterizing the final state of the three-body decay process. Note that this measurement is carried out one molecule at a time, at a rate of 10–100 three-body events per second.

In the following, we consider suitable projections of this high-dimensional set of information to gain insight into the fragmentation pattern. We introduce six parameters which uniquely describe the three fragment momentum vectors (c.f. Fig. 8.8). The $\mathbf{u}_i$ are contained in a plane. For each event, we define a new coordinate system (x', y', z') by the normal vector on this plane (z'-axis) and the direction of the largest momentum vector observed (x'-axis). Three Euler angles (ψ, θ, ϕ) describe the orientation of the (x', y', z')-coordinate system within the laboratory reference system (x, y, z), defined by the electric vector of the laser beam (z-axis) and the direction of the neutral beam (x-axis). The spatial orientation of the (x', y', z')-coordinate frame in Fig. 8.8 is determined by the spatial anisotropy of the photoexcitation process. For the remaining three parameters describing the arrangement of the three momenta in the (x', y')-plane we may choose the absolute values

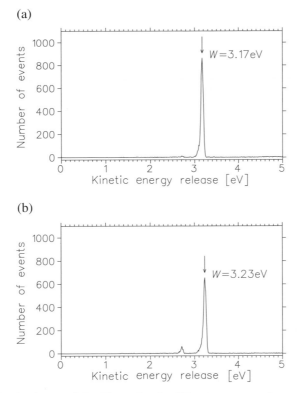

Fig. 8.9. Three-body decay of the H_3 molecule. Kinetic energy release spectra of the (a) $3s\ ^2A_1'\{v_1 = 0, v_2 = 0\}(N = 1, G = 0)$ and (b) $3d\ ^2E''\{v_1 = 0, v_2 = 0\}(N = 1, G = 0)$ states.

of the momenta $p_i = m|u_i|$ or the individual fragment energies $\epsilon_i = m\mathbf{u}_i^2/2$. We use the total kinetic energy W and two parameters showing the correlation among the fragment momenta.

The spectra of the total c.m. kinetic energy release $W = m(\mathbf{u}_1^2 + \mathbf{u}_2^2 + \mathbf{u}_3^2)/2$ pose a stringent test on the quality and precision of our data acquisition and reduction procedure. Since the energy of the laser-excited initial state above the three-body limit is laser-selected, W has to appear as a discrete observable. Fig. 8.9 shows the kinetic energy release spectrum from triple coincidence data of the H_3 $3s\ ^2A_1'$ and $3d\ ^2E''$ initial states binned at 10 meV resolution. In the spectrum of the $3s\ ^2A_1'$ ($N = 1$, $K = 0$) state, a narrow peak appears at 3.17 eV which is very close to the known energy of the initial state [18] above the three-body limit, 3.155 eV. The excellent agreement between measured *absolute* values of W and the previously known state energies confirms the quality of the absolute energy calibration of the detector. The kinetic energy spectrum of the $3d\ ^2E''$ state shows a strong peak at 3.23 eV which corresponds to break-up of the initial state into three hydrogen atoms, expected at a position of 3.230 eV. Additionally, an initially unexpected small peak is found at 2.71 eV. Events with this reduced value of W arise from three-body break-up following the radiative transition $3d\ ^2E'' \to 3p\ ^2E'$ [40] at 2.38 μm (see Fig. 8.4). No continuous background is observed in the kinetic energy release spectra of the three-body decay. This observation reveals that radiative transitions from the $3s\ ^2A_1'$ and $3d\ ^2E''$ states to the repulsive ground state lead to $H + H_2$ fragment pairs as the only exit channel [19]. It also indicates that the previously observed continuous photoemission spectra [49,50] are accompanied by two-body decay only.

8.6 Presentation of high-dimensional imaging data

To show the correlation among the fragment momenta within the (x', y') plane, we use a Dalitz plot [51]. For each event, we plot $(\epsilon_3/W - 1/3)$ versus $((\epsilon_2 - \epsilon_1)/(W \cdot \sqrt{3}))$. Energy and momentum conservation require that the data points in this Dalitz plot lie inside a circle with radius $1/3$, centred at the origin. In a Dalitz plot, the phase space density is conserved which means that a fragmentation process with a matrix element independent of the configuration leads to a homogeneous distribution in the kinematically allowed region. Preferred fragmention pathways can immediately be recognized from the event density in such a plot.

In Fig. 8.10a and b, the triple-coincident events following three-body break-up of the laser-excited H_3 $3s\ ^2A_1'$ and $3d\ ^2E''$ states are shown in Dalitz plots, selecting only events within the narrow kinetic energy regions around the strong peaks in Fig. 8.9 (3.05 to 3.25 eV and 3.1 to 3.35 eV respectively). Since the three hydrogen atoms are indistinguishable, points are drawn in Fig. 8.10 for the six permutations of the fragment energies ϵ_i measured in each event. In Fig. 8.10c, the correspondence between the configuration of the fragment momenta and the location in the plot is visualized. The three-fold rotation symmetry around the origin and the mirror symmetry with respect to the dashed lines in Fig. 8.10c result from the equal masses of the fragments. In order to understand the meaning of the very pronounced islands of correlation appearing in the experimental data in Fig. 8.10a and b, the limited collection efficiency of the detector has to be discussed first. Currently, detection of fragment triples where one of the momenta is close to zero (linear configuration) and therefore hits the space between the two detectors is excluded from detection as a three-body event. Also, fragment hits which are too close in time as well as in spatial coordinates (H + H-H configuration) are suppressed due to the finite pulse pair resolution of 10 ns. The geometric and electronic detector collection efficiency was determined in a Monte Carlo simulation by generating a uniform distribution of fragmentation configurations and calculating the fragment propagation to the detector. A Monte Carlo simulation of a typical detector response is shown in Fig. 8.10d for the $3d$ state. The collection efficiency vanishes only for the linear

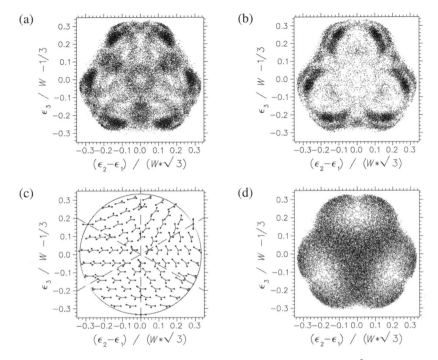

Fig. 8.10. Dalitz plot of the raw data from three-body decay of the $3s\ ^2A_1'$ (a) and $3d\ ^2E''$ (b) initial states of H_3. In (c), the correspondence between the location in the plot and the fragmentation configuration is indicated. In (d) a Monte Carlo simulation for a random distribution of fragmentation configuration in the phase space is shown, demonstrating the effect of the geometric detector collection efficiency.

and the $H + H_2$ configurations on the circle boundary. The remainder of the Dalitz plot area shows a smooth variation of the detection efficiency.

8.7 Discussion of experimental data

As a consequence of the Monte Carlo results, we are able to attribute the islands of high point density in Fig. 8.10a and b to the correlation of the fragment momenta produced by the dissociation process itself. We can definitely exclude experimental artefacts as contributing to this patterning. Nevertheless, we correct the measured distributions by weighting the data points with the geometric collection efficiency. In Fig. 8.11, the resulting three-body final state distributions of the $3s\ ^2A_1'$ and $3d\ ^2E''$ states are shown in grey scale images. Despite the high symmetry (D_{3h}) of the initial molecular states, asymmetric fragmentation configurations are very much preferred in finding a path into the three-particle continuum. Neither the totally symmetric configuration (centre of the plot) nor isosceles configurations (dashed lines in Fig. 8.10c show preferred population. It may be a surprise that the preferred

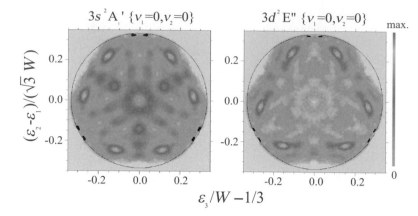

Fig. 8.11. Dalitz plots of the three-body decay of the $3s\ ^2A_1'$ (a) and $3d\ ^2E''$ (b) initial states of H_3. The geometric collection efficiency was calculated by a Monte Carlo simulation and the measured data were corrected.

fragmentation configurations sensitively depend on the initial state, although the absolute energies of the states investigated here differ as little as 75 meV and the nuclear equilibrium configuration for each initial state is extremely close to that of the vibrationless H_3^+ ion in its ground electronic state and its lowest rotational level. The striking difference of the final state distributions reflects the different coupling mechanism active between the initial state and the two sheets [49] of the repulsive ground state potential energy surface. The same argument should hold for the marked differences in the final state distributions of the two-body decay shown in Fig. 8.6. We observe the general tendency, that the two-body as well as the three-body distributions of the $3s\ ^2A_1'$ state show extremely rich structures with numerous minima and maxima. The distributions of the $3d\ ^2E''$ state are more concentrated on a few features. The suggestion is that in the two cases discussed here the zero point vibration, while exploring practically identical regions of internuclear configuration in the two cases considered here, finds individual regions of coupling to the continuum which are specific to each electronic state.

From a symmetry point of view, the breakdown of the Born–Oppenheimer approximation for the $3s\ ^2A_1'$ state is mediated by the zero-point motion in the degenerate vibration. In case of the $3d\ ^2E''$ state, the coupling is induced by the rotational tumbling motion [20] which is much slower than the vibrational motion. Therefore, the lifetime of the $3d\ ^2E''$ state is much longer than the $3s\ ^2A_1'$ state. In a semi-classical picture, the vibrational motion explores on a very short time scale many geometries where transitions between the initial state and the ground state surface may occur. Trajectories starting from these geometries interfere producing rich patterns in the final state. In the case of rotational coupling, only a few entrance points to the ground state surface are apparently significant and the final state distributions are governed by the trajectories starting from these points. In a quantum mechanical picture, the vibrational coupling matrix elements induce additional lobe structures in the nuclear wave function which then starts its time evolution on the repulsive surface. In the case of the rotational coupling, the wave function which enters the dynamics is to a good approximation the vibrational wave function of the initial state which has only one maximum at the equilibrium geometry.

While the coupling between the bound states and the continuum mediates the first entry of the quasi-bound system into the continuum, a series of avoided crossings between the upper sheet of the repulsive ground state surface and the s- and d-Rydberg states of $^2\Sigma_g^+$ symmetry in linear geometry (Petsalakis *et al.* [52]) will govern the further evolution of the continuum state.

8.8 Summary

We have demonstrated that fast beam translational spectroscopy is a powerful method to investigate laser-photofragmentation of molecular systems into neutral products. Recent progress in detector performance has opened up the avenue to study break-up processes into three and more products. The detection of the fragments in coincidence and the accurate measurement of their impact positions and arrival time differences allows us to determine the vectorial fragment momenta in the centre of mass frame. Studies which combine quantum mechanically complete preparation of the initial state and kinematically complete final state analysis in a photodissociation experiment have become feasible.

As an example to demonstrate the performance of the method we have investigated the dissociation of laser-excited H_3 molecules. We observe not only two-body decay into $H + H_2(v, J)$ fragment pairs but also a significant contribution of three-body decay into neutral hydrogen atoms. In the case of two-body decay, we have determined the rovibrational distribution of the $H_2(v, J)$ fragment. In the case of three-body decay, we measure the final state distribution in the kinematically accessible phase space and observe rich momentum correlation structures. The final state vibrational distributions of the H_2 fragment agree comparatively well with the results of a two-dimensional wave packet calculation. Not much theory exists on the three-body decay except for a prediction of the two-body/three-body branching ratios [47]. The detailed maps of momentum partition among the three fragments obtained in our experiment pose a significant challenge to quantum-structure and quantum-dynamics calculations.

Acknowledgements

This work was made possible by the generous support by Deutsche Forschungsgemeinschaft (SFB 276, TP C13).

References

1. T. Oka. In *Dissociative Recombination: Theory, Experiments and Applications IV*, eds. M. Larsson, J. B. A. Mitchell, I. F. Schneider, (World Scientific, Singapore, 2000), pp. 13–24.
2. U. Müller, Th. Eckert, M. Braun, H. Helm, *Phys. Rev. Lett.* **83**, 2718 (1999).
3. C. Maul, T. Haas, K. H. Gericke, *J. Phys. Chem.* **101**, 6619 (1997); C. Maul, K. H. Gericke, *Int. Rev. Phys. Chem.* **16**, 1 (1997).
4. Y. Tanaka, M. Kawaski, Y. Matsumi, H. Futsiwara, T. Ishiwata, L. J. Rogers, *et. al.*, *J. Chem. Phys.* **109**, 1315 (1998).
5. J. J. Lin, D. W. Hwang, Y. T. Lee, X. Yang, *J. Chem. Phys.* **108**, 10061 (1998).
6. M. Lange, O. Pfaff, U. Müller, R. Brenn, *Chem. Phys.* **230**, 117 (1998).
7. K. Wilson, *Disc. Farad. Soc.* **44**, 234 (1967).
8. J. T. Moseley, P. C. Cosby, J.-B. Ozenne, J. Durup, *J. Chem. Phys.* **70**, 1474 (1979).
9. H. Helm. Ion Interactions Probed by Photofragment Spectroscopy. In *Electronic and Atomic Collisions*, Inv. Papers of the XII ICPEAC, Berlin, (North Holland, Amsterdam, 1984), pp. 275–93.
10. H. Helm, D. DeBruijn, J. Los, *Phys. Rev. Lett.* **53**, 1642 (1984).
11. A. B. van der Kamp, P. B. Athmer, R. S. Hiemstra, J. R. Peterson, W. J. van der Zande, *Chem. Phys.* **193**, 181 (1995); L. D. A. Siebbeles, E. R. Wouters, W. J. van der Zande, *Phys. Rev. A* **54**, 531 (1996); E. A. Wouters, L. D. A. Siebbeles, P. C. Schuddeboom, B. R. Chalamala, W. J. van der Zande, *Phys. Rev. A* **54**, 522 (1996); B. Buijsse, E. R. Wouters, W. J. van der Zande, *Phys. Rev. Lett.* **77**, 243 (1996); E. R. Wouters, B. Buijsse, J. Los, W. J. van der Zande, *J. Chem. Phys.* **106**, 3974 (1997).
12. L. D. Gardner, M. M. Graff, J. L. Kohl, *Rev. Sci. Instrum.* **57**, 177 (1986).
13. R. E. Continetti, D. R. Cyr, D. L. Osborn, D. J. Leahy, D. M. Neumark, *J. Chem. Phys.* **99**, 2616 (1993).
14. K. A. Hanold, A. K. Luong, R. E. Continetti, *J. Chem. Phys.* **109**, 9215 (1998).
15. K. A. Hanold, A. K. Luong, T. G. Clements, R. E. Continetti, *Rev. Sci. Instrum.* **70**, 2268 (1999).
16. H. Helm, P. C. Cosby, *J. Chem. Phys.* **90**, 4208 (1989).
17. C. W. Walter, P. C. Cosby, H. Helm, *J. Chem. Phys.* **99**, 3553 (1993).
18. P. C. Cosby, H. Helm, *Phys. Rev. Lett.* **61**, 298 (1988).
19. U. Müller, P. C. Cosby, *J. Chem. Phys.* **105**, 3532, (1996).
20. U. Müller, P. C. Cosby, *Phys. Rev. A* **59**, 3632 (1999).
21. M. Braun, M. Beckert, U. Müller, *Rev. Sci. Instrum.* **71**, 4535 (2000).
22. M. Beckert, U. Müller, *Eur. Phys. J. D* **12**, 303 (2000).
23. D. P. de Bruin, J. Los, *Rev. Sci. Instrum.* **53**, 1020 (1982).
24. H. Helm, P. C. Cosby, *J. Chem. Phys.* **86**, 6813 (1987).
25. S. E. Sobottka, M. B. Williams, *IEEE Trans. Nucl. Sci.* **35**, 348 (1988).
26. K. Beckord, J. Becker, U. Werner, H. O. Lutz, *J. Phys. B* **27**, L585 (1994).
27. Z. Amitay, D. Zajfman, *Rev. Sci. Instrum.* **68**, 1387 (1997).
28. O. Jagutzki, V. Mergel, K. Ullmann-Pfleger, L. Spielberger, U. Meyer, H. Schmidt-Böcking, SPIE Proc. 'Imaging Spectroscopy IV', San Diego 19.7.–24.7.1998, in print.
29. R. Moshammer, J. Ullrich, M. Unverzagt, W. Schmitt, P. Jardin, R. E. Olson, *et al.*, *Nucl. Instrum. Methods. A* **108**, 425 (1996).
30. R. Doerner, V. Mergel, L. Spielberger, M. Achler, Kh. Khayyat, T. Vogt, *et al.*, *Nucl. Instrum. Methods. B* **124**, 225 (1997).
31. Z. Vager, E. P. Kanter, G. Both, P. J. Cooney, A. Faibis, W. Koenig, *et al.*, *Phys. Rev. Lett.* **57**, 2793 (1986).

32. D. Kella, M. Algranati, H. Feldman, O. Heber, H. Kovner, E. Malkin, *et al.*, *Nucl. Instrum. Methods. A* **329**, 440 (1990).
33. E. P. Kanter, Z. Vager, G. Both, P. J. Cooney, A. Faibis, W. Koenig, *et al.*, *Nucl. Instrum. Methods. B* **24/25**, 321 (1987).
34. D. Zajfman, E. P. Kanter, T. Graber, Z. Vager, R. Naaman, *Nucl. Instrum. Methods. B* **67**, 22 (1992).
35. O. Heber, D. Zajfman, D. Kella, Z. Vager, R. L. Watson, V. Horvat, *Nucl. Instrum. Methods. B* **99**, 90 (1995).
36. S. H. Courtney, W. L. Wilson, *Rev. Sci. Instrum.* **62**, 2100 (1991).
37. G. Leclerc, J.-B. Ozenne, J.-P. Corbeil, L. Sanche, *Rev. Sci. Instrum.* **62**, 2997 (1991).
38. T. Mizogawa, M. Sato, M. Yoshino, Y. Itoh, Y. Awaya, *Nucl. Instrum. Methods. A* **387**, 395 (1997).
39. C. Martin, P. Jelinsky, M. Lampton, R. F. Malina, H. O. Anger, *Rev. Sci. Instrum.* **52**, 1067 (1981).
40. G. Herzberg, J. T. Hougen, J. K. G. Watson, *Can. J. Phys.* **60**, 1261 (1982).
41. H. Helm, *Phys. Rev. Lett.* **56**, 42 (1986).
42. H. Helm, *Phys. Rev. A* **38**, 3425 (1988).
43. H. Helm, L. J. Lembo, P. C. Cosby, D. L. Huestis, *Fundamentals of Laser Interaction II*, Lecture Notes in Physics, ed. F. Ehlotzky, (Springer Verlag, Berlin, 1989), p. 264.
44. R. Reichle, I. Mistrík, U. Müller, H. Helm, *Phys. Rev. A*, **60**, 3929 (1999).
45. I. Mistrík, R. Reichle, U. Müller, H. Helm, M. Jungen, J. A. Stephens, *Phys. Rev. A*, **61**, 033410 (2000).
46. J. T. Hougen, *J. Chem. Phys.* **37**, 1433 (1962).
47. A. E. Orel, K. C. Kulander, *J. Chem. Phys.* **91**, 6086 (1989); J. L. Krause, K. C. Kulander, J. C. Light, A. E. Orel, *J. Chem. Phys.* **96**, 4283 (1992); J. L. Krause, A. E. Orel, B. H. Lengsfield, K. C. Kulander. In *Time-Dependent Quantum Molecular Dynamics*, eds. J. Broeckhove, L. Lathouwers, (Plenum, New York, 1992), p.131.
48. I. F. Schneider, A. E. Orel, *J. Chem. Phys.* **111**, 5873 (1999).
49. R. Bruckmeier, Ch. Wunderlich, H. Figger, *Phys. Rev. Lett.* **72**, 2550 (1994).
50. A. B. Raksit, R. F. Porter, W. P. Garver, J. J. Leventhal, *Phys. Rev. Lett.* **55**, 378 (1985).
51. R. H. Dalitz, *Philos. Mag.* **44**, 1068 (1953); *Ann. Rev. Nucl. Sci.* **13**, 339 (1963).
52. I. D. Petsalakis, G. Theodorakopoulos, J. S. Wright, *J. Chem. Phys.* **89**, 6850 (1988).

9

Collisions of HCl with molecular colliders at $\sim 540\,\mathrm{cm}^{-1}$ collision energy

ELISABETH A. WADE, K. THOMAS LORENZ,
JAMES L. SPRINGFIELD AND DAVID W. CHANDLER

9.1 Introduction

The use of spectroscopic techniques to study crossed molecular beam scattering has been both fruitful and difficult to accomplish. The advantage of using laser spectroscopy to detect the products in crossed molecular beam scattering experiments is that one can obtain quantum-state-selective information about the scattering process. The disadvantage is that it is difficult to design an experiment with sufficient wavelength and spatial resolution and detection sensitivity. For this reason until quite recently laser induced fluorescence (LIF) was the only laser-based technique used for the quantum-state-selective detection of scattering products [1–5]. Recently the technique of ion imaging has been used to increase the sensitivity of ionization detection such that resonance-enhanced multi-photon ionization (REMPI) can now be used to detect molecular beam scattering products. Suits *et al.* were the first, in the early 1990s, to apply ion-imaging techniques to bimolecular scattering [6]. Since then, ion imaging has been found to be a powerful tool for the study of bimolecular inelastic scattering. Ion imaging has been used to measure differential cross-sections (DCSs) [6,7], as well as to measure collision-induced rotational alignment [8] and orientation [9]. In this chapter we will focus on a new application of ion imaging, the retrieving of correlated energy transfer distributions from crossed molecular beam ion imaging experiments.

The previous studies of bimolecular collision systems consisted of a diatomic target molecule colliding with a rare gas atom. The monatomic collider gas has no internal energy, and a single rotational state of the diatomic molecule was detected, using REMPI. Each image, therefore, corresponded to a single ring we refer to as a 'scattering sphere' whose intensity pattern is a function of the DCS for that quantum state and the detection sensitivity for each scattering angle. The size of the scattering sphere is dictated by conservation of energy and momentum, the Newton spheres. In this study, the collider gas is itself molecular, either N_2 or CH_4, and so

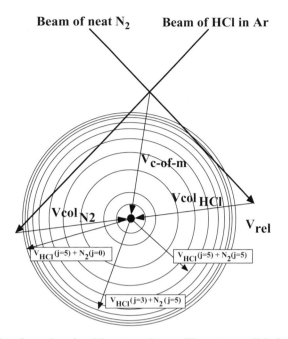

Fig. 9.1. Schematic of overlapping Newton spheres. The outer, solid circle represents HCl ($j_{HCl} = 4$) correlated with N_2 ($j_{N2} = 0$), and the inner, grey circles represent HCl ($j_{HCl} = 4$) correlated with N_2 ($j_{N2} = 2, 4, 6, 7, 8, 9, 10$ and 11). Initially, all j_{N2} states are not shown because they are nearly coincident and would not be resolvable. The $\mathbf{v}_{rel}$ vector shows the 'tilt' in the image resulting from the different velocities of HCl (seeded in Ar) and N_2.

has its own internal degrees of freedom. Figure 9.1 displays the Newton diagram for the HCl ($j = 0$) + N_2 $\rightarrow$ HCl ($j = 4$) + N_2 ($j = 0 \ldots \ldots 11$) reaction showing the individual N_2 Newton spheres correlating to the particular quantum state of the HCl. Each observed product channel, indicated by an individual Newton sphere, can have a unique DCS which leads to a unique intensity pattern in the scattering. The DCS is a function of the rotational states of both products and the potential energy surface for the collision. A given inelastic collision can change the internal states of both the target molecule (HCl) and the collider molecule. Only the HCl molecules populating a particular rotational state are detected, and these HCl molecules are associated with a set of final rotational states of the collider (N_2 or CH_4) molecule. The final ion image, therefore, is a superposition of concentric scattering spheres, centred about the centre of mass for the particular scattering pair, with each ring corresponding to a pair of correlated product rotational states.

If the velocity spreads in the molecular beams were sufficiently narrow, we would observe a series of concentric rings when the scattering spheres were compressed into an ion image. The relative intensities of these rings provides information about the relative integrated cross-section as a function of Δj for the collider molecule.

The intensity pattern within each ring gives information about the differential cross-section for that pair of final rotational states. However, in these experiments the velocity spreads in the molecular beams are significant, $\sim 10\%$, the rotational states of the N_2 and CH_4 are closely spaced in energy, and the projection of the ions spheres onto the detector causes different channels to overlap. Therefore, in our data the concentric scattering rings associated with the rotational states of the N_2 or CH_4 overlap with their neighbors. This makes it difficult to extract the information about individual DCSs and correlated energy distributions cleanly and unambiguously. However, we have extracted this information with reasonable error bars. The larger the rotational constant of the collider, the greater the spacing between the Newton spheres, and the better the chance of extracting information from the overlapping rings. N_2 has a small rotational constant, 2.010 cm^{-1} [10]. CH_4 is a symmetric top, with a rotational constant of 5.2412 cm^{-1} [11].

The colliders were chosen for their molecular masses, because the mass of CH_4 is nearly that of Ne and N_2 is between that of Ne and Ar. The differences in collision behavior for HCl with the rare gas colliders Ne, Ar, and Kr have been found to be primarily due to differences in kinematics, while the scaled potential energy surfaces are nearly identical [12]. Since the kinematics for $HCl + N_2$ and $HCl + CH_4$ are comparable to the kinematics for $HCl + Ne$ and $HCl + Ar$, any differences between the observed ion images must be due to differences in the potential energy surface, not simple kinematic differences.

9.2 Experimental

The experimental apparatus has been described previously [7]. In these experiments, HCl (seeded 2–5% in Ar) is the target gas. The rotational temperature of the HCl in the beam is low enough that about 97% of the HCl is initially in the $j''_{HCl} = 0$ state and the remaining HCl is primarily in $j''_{HCl} = 1$. Neat N_2 (house nitrogen) and CH_4 (99.99%, Matheson) are used as collider gases. Using pure gases in the collider beam resulted in a hotter molecular beam for the collider than for the target gas. A $2 + 2$ REMPI spectrum is collected for N_2 [13], and the beam temperature is found to be $25 \pm 5 \text{ K}$, which means that the rotational states up to $j''_{N2} = 3$ are substantially populated before the collision. The rotational temperature of the neat CH_4 molecular beam, for which a REMPI scheme is not available, is assumed to be comparable. Very pure CH_4 is necessary because the ionization laser caused the hydrocarbon impurities in the CH_4 molecular beam to nonresonantly dissociate and ionize, which results in a large background signal that is associated with the collider beam and can not be easily subtracted. The collision energy is $541 \pm 70 \text{ cm}^{-1}$ for $HCl + N_2$ and $549 \pm 70 \text{ cm}^{-1}$ for $HCl + CH_4$. This is very close to the collision energy for $HCl + Ar$, $538 \pm 70 \text{ cm}^{-1}$ that was previously published.

HCl + Ar compared with molecular colliders

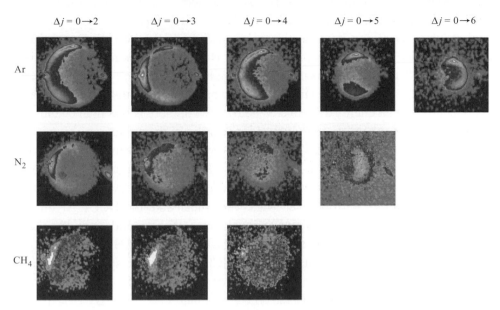

Fig. 9.2. Ion images for HCl + Ar, N_2, and CH_4. These images are rotated so that the center line is parallel to the top and bottom of the images.

9.3 Comparison of images of HCl + molecular colliders with HCl + Ar images

The ion images for HCl + N_2 and HCl + CH_4 are shown in Fig. 9.2, along with the previously published images of HCl + Ar. The target beam, containing the HCl, is travelling from the top right to bottom left in the image while the collider beam is travelling top left to bottom right (see Fig. 9.1). Ion images are necessarily symmetric about the velocity vector of the collision in the center-of-mass frame of reference, because for a given deflection angle, ϑ, the probability of being scattered at any azimuthal angle is equal. When HCl collides with Ar, the velocity of the target and collider beams in the laboratory frame are nearly identical. Therefore, the line of symmetry through the image, the relative velocity vector of the collision in the center-of-mass frame of reference, appears horizontal in the ion images. When the collider gas is lighter than the carrier gas, as are both N_2 and CH_4, the velocity of the collider beam is greater, and the line of symmetry is tilted so that the image appears to be rotated clockwise. The data in Fig. 9.2 have been rotated so that the central line, the collision velocity vector in the center-of-mass frame, is again horizontal in the image.

The HCl + N_2 and HCl + CH_4 images are substantially different from the HCl + Ar images. First of all, the total signal (and therefore the integrated cross-section) for these molecular colliders is much smaller than for HCl + Ar. The integrated cross-section, for a given j_{HCl}, is estimated to be 10–20 times smaller for HCl + N_2, and 100–500 times smaller for HCl + CH_4, based on observed signal and required averaging time. Also, ion images for the highest rotational states, $j_{HCl} = 6$ for HCl + N_2 and $j_{HCl} = 5$ and 6 for HCl + CH_4, could not be observed. It is possible that the collisions that result in high j_{HCl} also result in high Δj for the molecular collider, and that when both collision partners are in high rotational states, there is not enough energy available. When $j_{HCl} = 6$, there is less than 60 cm^{-1} of energy available either for translation or for rotation of the collider.

The implication that high Δj_{HCl} is associated with high Δj for the collider, while low Δj_{HCl} is associated with low Δj for the collider, is also supported by qualitative observation of the other images. For $j_{HCl} = 2$ and 3, the HCl + N_2 and HCl + CH_4 images have intense, narrow outer rings, very much like the HCl + Ar images, which suggests that relatively few rotational states of the collider are populated. For $j_{HCl} \geq 4$, on the other hand, the centre of the HCl + N_2 and HCl + CH_4 images are much more filled in than the equivalent HCl + Ar images, suggesting that for high Δj_{HCl} many rotational states of the collider are populated.

The HCl + N_2 images resemble the HCl + Ar images for $\Delta j = \Delta j + 1$. In other words, the HCl ($j_{HCl} = 2$) + N_2 ion image resembles the HCl ($j_{HCl} = 3$) + Ar image, the HCl ($j_{HCl} = 3$) + N_2 image resembles the HCl ($j_{HCl} = 4$) + N_2 image, etc. The HCl + CH_4 ion images are strongly forward-scattered for $j_{HCl} = 2$ and 3. This is why the HCl ($j_{HCl} = 2, 3$) + CH_4 images appear to have such poor signal to noise: at least 80% of the signal is concentrated in the first 20°.

9.4 Differential cross-sections from ion images of HCl + $N_2(\Delta j < 5)$ and HCl + $CH_4(\Delta j < 4)$

The DCSs for HCl + $N_2(\Delta j < 5)$ and HCl + $CH_4(\Delta j < 4)$ have been extracted, and are compared with the DCS for HCl + Ar in Fig. 9.3. The DCSs are extracted from the ion images using an interative procedure that has been described in detail elsewhere [7], and will be summarized here. First, an isotropic DCS, DCS$(\vartheta) = 1$, is input into a scattering simulation program which produces a simulated image that incorporates the apparatus function, including detectivity biases and the molecular beam spreads. Then, an annulus is defined around the rim of the experimental and simulated images, and used to determine two angular intensity distributions. The experimental angular intensity distribution is divided by the simulated angular intensity distribution to produce the approximated DCS. The approximated DCS is then used in the scattering simulation program to produce a new simulated image and a new angular intensity distribution, and the experimental angular intensity distribution is divided by the new simulated angular intensity distribution. This

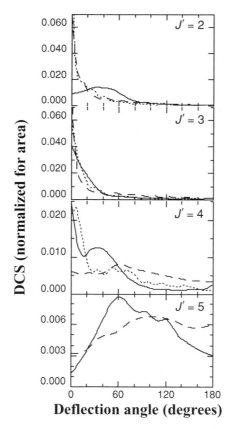

Fig. 9.3. Extracted differential cross-sections from the outer edges of the images of Fig. 9.2. The outer edge correlates with low J target quantum state. The solid line is HCl + Ar, the dashed line is HCl + N_2, and the dotted line is HCl + CH_4.

gives a correction function which is multiplied by the first approximated DCS to give a new DCS. This process continues until successive DCSs converge. This iterative procedure pulls out sharp features at very low and very high deflection angles, which tend to wash out.

The extracted DCSs displayed in Fig. 9.3 are based on annuli set on the outer ring of the ion images, which corresponds to $\Delta j_{N2} < 5$ and $\Delta j_{CH4} < 4$. The solid line is HCl + Ar, the dashed line is HCl + N_2, and the dotted line is HCl + CH_4. The trends observed in the ion images can also be observed here. For low Δj_{HCl} ($j_{HCl} = 2$ and 3) the molecular colliders resulted in DCSs that are more forward-scattered than were observed for HCl + Ar, while for high Δj_{HCl} ($j_{HCl} = 4$ and 5) the molecular colliders resulted in DCSs that are more back-scattered.

These differences are substantial and cannot be accounted for by the differences in kinematics [12]. The implications of the HCl + CH_4 DCSs are particularly interesting, since the HCl is strongly forward-scattered when the methane has little internal energy (outer rim of image) even for $j_{HCl} = 4$. While the internal states of CH_4 will complicate the analysis, the outer edge, where $\Delta j_{CH4} < 4$, might be expected to show the same behavior as a monatomic collider, since CH_4 is a symmetric top. Instead, the DCSs suggest that, even for relatively large Δj_{HCl} and low Δj_{CH4}, HCl and CH_4 collisions are glancing, and do not result in large deflection angles. The HCl + N_2 DCSs are more difficult to interpret, but also suggest that the potential energy surface for HCl + N_2 is distinctly different than that of HCl + rare gases.

9.5 Extracting DCSs from HCl($j_{HCl} = 5$) and N_2

By using the iterative method described above we can see how the differential cross-section for a particular state of HCl changes with internal energy of the collision partner. We will extract multiple DCSs from the HCl($j_{HCl} = 5$) + N_2 image, even though multiple rings are not clearly visible. At first glance the HCl($j_{HCl} = 5$) + N_2 image appears to be very similar to the HCl($j_{HCl} = 6$) + Ar image, but when a circle representing the outer edge of the scattering sphere is overlaid on both images, as in Fig. 9.4, an important difference can be observed. The image of HCl($j_{HCl} = 6$) + Ar is highly back-scattered, but is also clearly circular. The image of HCl($j_{HCl} = 6$) + N_2, on the other hand, is not circular but is nearly a vertical stripe. This stripe results from the overlap of multiple scattering spheres with different DCSs. This effect can be seen even more clearly when a second, inner circle is overlaid on the image. Inside the second circle, the angular intensity distribution for HCl($j_{HCl} = 6$) + Ar is the same as the angular intensity distribution between the two circles. For HCl($j_{HCl} = 5$) + N_2, the equivalent angular

Multiple DCSs from molecular colliders

Ar, $j' = 6$ N_2, $j' = 5$

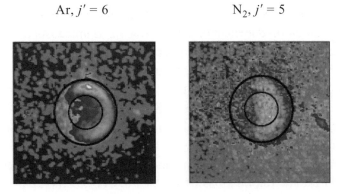

Fig. 9.4. Comparison of $j_{HCl} = 6$ for HCl + Ar and $j_{HCl} = 5$ for HCl + N_2.

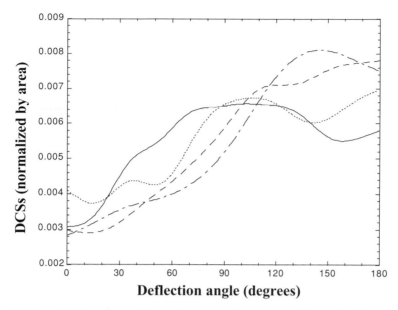

Fig. 9.5. Multiple DCSs extracted from the $j_{HCl} = 5$ image for HCl + N_2. The solid line is the DCS correlated with $\Delta j_{N2} \sim 0$. The dotted line is the DCS correlated with $\Delta j_{N2} \sim 2$. The dashed line is the DCS correlated with $\Delta j_{N2} \sim 5$. The dot-dashed line is the DCS correlated with $\Delta j_{N2} \sim 8$.

intensity distributions are clearly different, with the part of the image inside the inner circle clearly more back-scattered than the part of the image between the two circles.

Multiple DCSs extracted from the HCl($j_{HCl} = 5$) + N_2 image are shown in Fig. 9.5. The procedure used to extract these DCSs is a variation on the normal procedure used to extract DCSs from ion images involving atomic colliders [7]. First, a narrow annulus is defined along the outer rim of the ion image, and a DCS is extracted, using the procedure described above. This DCS is used to simulate an ion image, which is subtracted from the original experimental ion image. Then, a new annulus is defined along the rim of the resulting smaller image. The new, smaller annulus correlates to scattering into higher rotational states of N_2. This correlation is approximate and the DCS extracted can not be associated with a particular rotational state of N_2.

Because of signal-to-noise considerations and uncertainties in the iterative procedure for extracting successive DCSs, we will only look for trends in the data. The DCSs extracted from the HCl($j_{HCl} = 5$) + N_2 image are increasingly back-scattered as Δj_{N2} increases. An increase in back-scattering with rotational state is seen in many different systems [14], and appears to be the case here, within a single image.

9.6 Conclusions

We have observed and analyzed the correlated scattering patterns for collisions of HCl molecule with a diatomic collider, N_2, and a polyatomic collider, CH_4. We find that in general the scattering is of smaller intensity than the corresponding diatomic/atom collision. For HCl ($j_{HCl} = 2, 3$) being formed, we find that molecular colliders are more forward-scattered than atomic colliders. For HCl ($j_{HCl} = 4, 5$), however, we find the opposite trend. We find that the scattering tends to be more backward scattered than the corresponding diatomic/atom collision. This implies that in order to generate HCl in a highly rotationally excited quantum state with a diatomic or polyatomic collider, one needs to strike it with a smaller impact parameter compared to an atomic collision. We also note that high Δj_{HCl} appears to be correlated with high Δj in the collider, while low Δj_{HCl} appears to be correlated with low Δj in the collider.

Acknowledgement

This work was supported by the Department of Energy, Office of Basic Energy Sciences, Division of Chemical Sciences.

References

1. C. T. Rettner, L. Woste, R. N. Zare, *Chem. Phys.* **58**, 371 (1981).
2. G. Hall, C. F. Giese, W. R. Gentry, *J. Chem. Phys.* **83**, 5343 (1985).
3. D. J. Krajnovich, K. W. Butz, H. Du, C. S. Parmenter, *J. Chem. Phys.* **91**, 7705 (1989).
4. H. M. Keller, M. Kulz, R. Setzkorn, G. Z. He, K. Bergmann, H. G. Rubahan, *J. Chem. Phys.* **96**, 8817 (1992).
5. K. Schreel, J. J. ter Meulen, *J. Phys. Chem. A*, **101**, 7639 (1997).
6. A. G. Suits, L. S. Bontuyan, P. L. Houston, B. J. Whitaker, *J. Chem. Phys.* **96**, 8618 (1992).
7. K. T. Lorenz, M. S. Westley, D. W. Chandler, *Phys. Chem. Chem. Phys.* **2**, 481 (2000).
8. J. I. Cline, K. T. Lorenz, E. A. Wade, J. W. Barr, D. W. Chandler, *J. Chem. Phys.* **115**, 6277 (2001).
9. K. T. Lorenz, D. W. Chandler, J. W. Barr, W. Chen, G. L. Barnes, J. I. Cline, *Science* **293**, 2063 (2001).
10. G. Herzberg. In: *Molecular Spectra and Molecular Structure, I. Spectra of Diatomic Molecules*, (Kreiger, Malabar, 1989), p. 553.
11. G. Herzberg. In *Molecular Spectra and Molecular Structure, III. Electronic Spectra and Electronic Structure of Polyatomic Molecules*, (Kreiger, Malabar, 1989), p. 619.
12. E. A. Wade, K. T. Lorenz, J. L. Springfield, D. W. Chandler, *J. Phys. Chem. A*, in preparation.
13. G. O. Sitz, R. L. Farrow, *J. Chem. Phys.* **93**, 7883 (1990).
14. R. D. Levine, R. B. Bernstein, *Molecular Reaction Dynamics and Chemical Reactivity*, (Oxford University Press, Oxford, 1987).

10

Measurement of state-resolved differential cross-sections of bimolecular reactions using single beam velocity mapping

T. PETER RAKITZIS AND THEOFANIS N. KITSOPOULOS

10.1 Introduction

The dynamics of chemical reactions can be probed in detail by employing a method of experimentation that is sensitive to both reactant and product internal state energy distributions and velocities and correlations between these quantities. Two main scattering techniques have emerged to study bimolecular reactions of the form $A + BC \rightarrow AB + C$.

One is the crossed molecular beam technique [1], in which molecular beams of the reactants A and BC, with well-defined beam velocities, are intersected at some angle. The AB and/or C product velocity and angular distributions are then measured with a form of universal ionization (such as electron impact ionization) and time-of-flight mass spectrometry (TOF–MS). The universal ionization step is used because the product densities are usually too small to allow a form of state selection, such as that of laser ionization, which provides small detection volumes.

Another method is a single beam technique [2,3], in which reactants or their precursors are premixed and co-expanded through a pulsed nozzle thus forming a single molecular beam. A photolysis laser is used to generate hot-atom reactants that subsequently react with other molecules in the beam. After a suitable time delay (typically 50–500 ns) that allows a sufficient density of products to build up, a second (probe) laser is used to state-selectively probe the reaction products, which are then detected with a velocity-sensitive technique (either TOF–MS, or Doppler spectrometry).

The single and crossed beam techniques have different strengths and weaknesses. Crossed molecular beams afford much better resolution in the scattering angle (although blind regions sometimes exist, depending on the kinematics of the reaction), and only large cross-section reactions can be studied readily. The main advantage of the single beam technique is that the reactant densities in the beam are much higher, and so reactions with small cross-sections can also be studied.

Additionally, another important advantage is that state selection of the products ensures product state resolution, as well as the measurement of the product angular momentum distributions.

Until recently, the product velocity distributions from photoinitiated reactions in a single beam have been measured with the one-dimensional techniques of TOF–MS and laser-induced fluorescence (LIF) combined with Doppler spectroscopy. The disadvantage of these one-dimensional techniques is that the signals of the various velocity channels, as well as their angular distributions, are strongly coupled, and elaborate basis-set analyses or forward convolution techniques are necessary for the data analysis.

Two-dimensional imaging and velocity mapping of the products overcomes these disadvantages, as the complete three-dimensional product velocity distributions can be measured with the inverse-Abel transform [4], or slice imaging [5]. However, the traditional experimental set-up of ion imaging machines uses molecular beams (sampled far from the nozzle orifice) with densities that are too low for reaction products to be observed readily.

In this chapter we demonstrate the use of velocity mapping for determining state-selective differential cross-sections for bimolecular reactions, by making experimental modifications to allow higher beam densities in the reaction region, thereby combining the advantages of high beam densities (for studying reactions with small cross-sections) and two-dimensional velocity mapping (which allows the decoupling of the measurement of product speed and angular distributions). In particular, we investigate the reaction of ground state atomic chlorine with ethane producing hydrogen chloride and the ethyl radical:

$$Cl(^2P_{3/2}) + C_2H_6 \rightarrow HCl + C_2H_5$$

For the convenience of this demonstration, this reaction was chosen because of its large reaction cross-section. However, the HCl product signal was also very large, indicating that reactions with significantly smaller cross-sections (at least two orders of magnitude) can be studied using this method. This reaction has been studied by Kandel *et al.* [6–8] at the collision energy of 0.24 ± 0.03 eV using single-beam TOF Doppler measurements to extract the velocity distributions of state-selected HCl or C_2H_5 products. It was concluded that the C_2H_5 reaction product has almost no internal energy excitation and little effect on the dynamics of the reaction. HCl is produced mostly in its $J = 0$ and $J = 1$ state with smaller amounts formed in higher rotational states. The high product number densities are obtained in the beam as the laser interaction region is only a few millimetres from the nozzle orifice. Our experiment is performed at the collision energy of 0.36 eV. Velocity map images for the $HCl(v = 0, J = 1)$ product are measured from which speed and angular distributions are extracted.

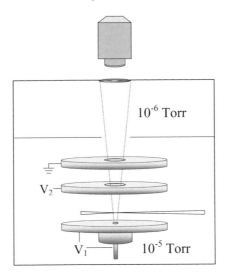

10^{-6} Torr

V_2

V_1 10^{-5} Torr

Fig. 10.1. Apparatus schematic. Co-propagating lasers intersect a pulsed molecular beam. The pump (photolysis) beam produces atomic chlorine that reacts with ethane. HCl product is state-selectively photoionized by the probe laser. The resulting ions are detected with a two-dimensional position sensitive imaging detector after passage through a linear time-of-flight mass spectrometer.

10.2 Experimental

A schematic of the experimental apparatus is shown in Fig. 10.1. The major modification to the apparatus compared with the configuration used for unimolecular photofragmentation studies [9] is the removal of the skimmer and mounting of the ion optics [10,11] directly in front of the nozzle. By doing so, differential pumping is being sacrificed for the increase of the number density of the reactants and consequently that of the reaction products. A slight differential pumping between the source and detector regions is achieved by a 15 mm hole. When a mixture of 20% Cl_2 (Merck, 99.8%) and 30% ethane (Aldrich, 99%) in He is expanded through the pulsed nozzle operating at 10 Hz, backing pressure ~ 1000 Torr, the source and detector chamber pressures are 10^{-5} and 10^{-6} Torr respectively. The repeller is mounted concentrically onto the nozzle faceplate and has a 2 mm orifice. Both nozzle and repeller are at the same voltage in order to eliminate discharging during the gas pulse firing.

Two co-propagating lasers intersect the molecular beam perpendicularly approximately 5–10 mm from the repeller plate surface. The photolysis laser beam is the unpolarized output (~ 50 mJ/pulse) of an Excimer laser (Lumonics HyperEX 400) operating at 308 nm, which photodissociates Cl_2. HCl($v = 0, J = 1$) product is detected via (2+1) REMPI through the $F^1\Delta_2$ state [12–14], using the frequency-doubled output of a Nd:YAG-pumped optical parametric oscillator

(Spectra-Physics Pro190, MOPO 730D10). Both laser beams are focused using the same 300 mm focal length lens. Because of the poor beam quality of the Excimer laser the photolysis laser beam dimensions at the focus are approximately 1 mm × 3 mm. The timing of the experiment is controlled using a digital delay pulse generator (SRS, DG535). The probe pulse is delayed by 200 ns with respect to the photolysis pulse, in order to allow a sufficient density of HCl product to build up prior to the REMPI detection. Background images are obtained by reducing this time delay to 30 ns, and these images are subtracted from the signal images.

The resultant ions are accelerated by a suitable electric field needed for velocity mapping and traverse a field-free TOF region prior to detection by a position-sensitive imaging detector consisting of a matched pair of MCP plates (Hamamatsu) coupled to a P20 phosphor anode (Proxitronic). The voltage applied to the phosphor anode is pulsed (0 to +3 kV), thus enabling mass gating in order to detect $H^{35}Cl^+$ ions only. The images are recorded using a charge-coupled device camera (MCD1000S, Spectra Source). The experiment is performed with sufficient care as to avoid space-charge effects.

10.3 Results and discussion

To extract quantitative data from the two-dimensional images one needs to reconstruct the three-dimensional (3-D) velocity distribution. For photodissociation experiments, in which linearly polarized lasers are commonly used, the configuration is such that the photolysis and probe polarization vectors are parallel to each other and to the imaging plane. The resulting photofragment angular distribution is cylindrically symmetric and thus an inverse Abel transform can be applied to extract the 3-D distribution [4]. In the present experiment, the photolysis laser is unpolarized, i.e., the electric field of the laser is cylindrically symmetric with respect to the propagation direction. Therefore, the resulting Cl-atom photofragment and the product HCl angular distributions will also be cylindrically symmetric with respect to this direction allowing us once again to apply the inverse Abel transformation to extract scattering information[1].

A velocity map image for the reactant Cl photofragment is presented in Fig. 10.2a. After the usual image processing [4,15,16] we determine the speed and angular distributions that are shown in Fig. 10.2b and c. From the speed distribution we estimate the speed resolution of our apparatus in this 'high-reactant-density' configuration is $\Delta v/v \approx 7\%$, approximately 2.5-times worse than the 2–3% resolution typically achieved in the 'low-reactant-density' photofragmentation experiments. The reason is purely the large interaction volume that is necessary to observe reaction

---

[1] We are ignoring possible alignment effects for now. The laboratory HCl alignment effects are small due to the Cl nuclear hyperfine depolarization.

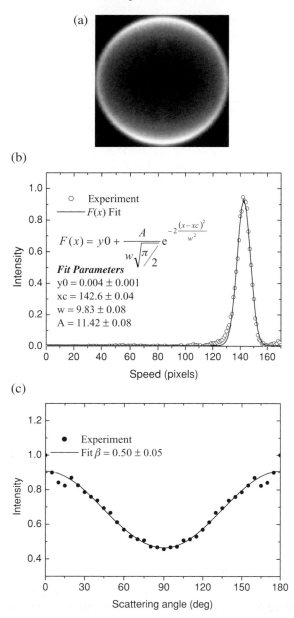

Fig. 10.2. (a) Velocity map image for the Cl-atom reactant. (b) Speed distribution for the Cl-atom reactant determined from the image in (a). The distribution is fitted using a Gaussian function whose parameters are shown in the figure. The speed resolution is determined from the ratio $\frac{w}{xc} \approx 7\%$. (c) Cl-atom reactant angular distribution determined from the image in (a). The anisotropy parameter is $\beta = 0.5 \pm 0.05$ as expected (see text).

products readily. The obtained scattering-angle resolution is at least as good as that of Kandel *et al.* The $Cl + C_2H_6$ reactive cross-section is large enough to be probed in the 'low-reactant-density' configuration as well, with the improvement in resolution mentioned above. The current experiment demonstrates that reasonable resolution can be obtained in the 'high-reactant-density' configuration necessary for reactions with small cross-sections.

The photolysis laser wavelength at 308 nm defines the speed of Cl reactants and hence the collision energy of the reaction. At this wavelength essentially only the ground spin-orbit state of Cl is produced, and the speed of Cl fragments after photodissociation is 2070 m/s. This reactant speed dictates a collision energy of 0.36 eV and a maximum allowed speed of 2300 m/s for HCl ($v = 0, J = 1$) products (assuming no internal excitation in the ethyl radical). For the calculation of the above values we have used 4.43 eV for the bond dissociation energy of HCl [17] and 4.31 eV for the C–H bond dissociation energy in ethane [18].

The uncertainty in the collision energy for the reaction in the current experimental set-up is determined by several factors. One consideration is the bandwidth of the photolysis laser which in this case is ~ 0.013 eV (XeCl Excimer). That means that the error in the photofragment kinetic energy is ± 6.5 meV. If a narrow band laser is used for the photolysis (0.2 cm^{-1}), then this uncertainty becomes negligible. Other factors are the velocity slip between the Cl_2 precursor reactant molecule and the ethane molecules and the molecular beam speed ratio, which are estimated to be about 5% [19].

The angular distribution for the Cl photofragment is shown in Fig. 10.2c. We notice a substantial discrepancy from the expected $[1 + \beta P_2(\cos\theta)]$ behaviour near the symmetry axis. This is partly caused by the Abel inversion procedure and more so in this case by a dark region on our detector. The distribution is anisotropic with a preferred direction along the laser propagation direction (symmetry axis). The anisotropy parameter that best describes the experimental data is $\beta(Cl) = 0.50 \pm 0.05$. For linearly polarized light, the photolysis of Cl_2 at 308 nm is described by $\beta = -1$ [11]. When unpolarized light is used we expect a change in the anisotropy parameter by a factor of -0.5 [20], i.e., we expect $\beta(Cl) = +0.5$, consistent with our experimental observations.

In the data image for HCl ($v = 0, J = 1$) shown in Fig. 10.3a, two main features are observed: a bright central spot and a broad anisotropic distribution. Figure 10.3a has been quadrant-averaged with respect to the photolysis laser propagation direction and the axis perpendicular to it. Performing the experiment under the same experimental conditions but without the photolysis laser at 308 nm and probing $H^{35}Cl$ ($v = 0, J = 1$), we still observe the central peak in the image but the surrounding anisotropic distribution is completely gone. This suggests that the 308 nm laser is involved in the production of the translationally energetic species observed

(a) (b)

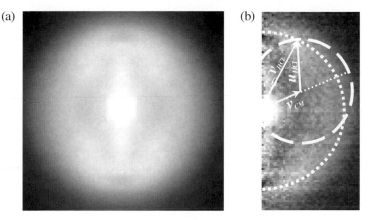

Fig. 10.3. (a) Velocity map image for the HCl($v = 0, J = 1$) reaction product. The broad energetic outer ring is the HCl reaction product, whereas the central spot is produced from residual HCl in the parent molecular beam. (b) Reconstructed images for the HCl($v = 0, J = 1$) reaction product using the inverse Abel transformation. Arrow indicates axis of symmetry. Overlaid are a Newton *Circle* (- - - -) (see text) for the HCl($v = 0, J = 1$) product, the CM laboratory velocity $\mathbf{v_{CM}}$, and two circles ($\cdots\cdots$) indicating the laboratory recoil speed $\mathbf{v_{HCl}}$ for forward ($\theta = 0$) scattering and scattering at some arbitrary angle θ. (c) Anisotropy parameter for the various $\mathbf{v_{HCl}}$ values. Curve A describes β when all the available energy to the ethyl radical is kinetic, and curve B describes β when all the available energy is deposited in the internal modes of the ethyl radical. We notice that for limiting values of $\mathbf{v_{HCl}}$, the β values approach those of the Cl-atom reactant (0.5), whereas for intermediate $\mathbf{v_{HCl}}$ values, i.e., $\mathbf{v_{HCl}} \approx \mathbf{v_{CM}}$, the β values change sign. (d) The differential cross-section for the HCl($v = 0, J = 1$) reaction product determined from the Abel-inverted image (b) using (10.2). Also shown is the HCl($v = 0, J = 1$) differential cross-section at 0.24 eV from Ref. [6].

in the outer portion of the image. We therefore conclude that the central spot is from residual HCl present in the molecular beam prior to the photolytic initiation of the reaction most probably formed by reactions in the mixing tank or the tubing that connects it to the nozzle.

Approximately 10 probe laser frequency steps are needed to scan the Doppler profile of the HCl ($v = 0, J = 1$) reaction product. Our signal to noise is sufficiently high that only 30 s signal averaging (300 laser shots) per step is needed for acquiring a useful image. In between each frequency step, the gas manifold is pumped out and flushed for about 10 s with the 5% NO/He mix. This procedure reduces the central spot such that it doesn't interfere with our data analysis. The exact mechanism of the HCl background production and why our empirical titration with NO remedies this problem remains unclear. Data acquisition at all frequency steps required about 10 min. The frequency-stepped images are co-added without laser-power normalization. The analysis of different images acquired this way is in close agreement.

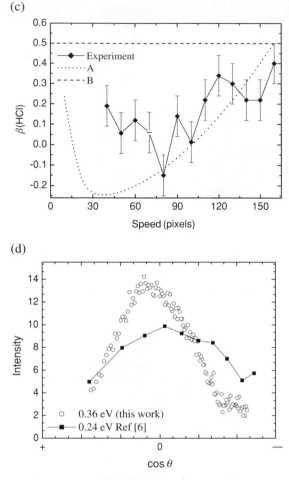

Fig. 10.3 (*cont.*).

Analysis of the images, to yield the differential cross-section, can be performed in two very similar ways (at first we will assume that there is no energy deposited in the ethyl fragment). The Abel-inverted images can be analyzed in the standard way to yield the speed distribution, by integrating the Abel-inverted image over all angles as a function of speed (radius r), and weighing the intensity of each pixel of the image by $r\sin\theta$ (surface element). To give the differential cross-section, the speed distribution is weighted by $(1/v)$, and the velocity axis is transformed to $\cos\theta$ by the relationship [21]:

$$\cos\theta = \frac{v_{HCl}^2 - u_{HCl}^2 - v_{CM}^2}{2|u_{HCl}||v_{CM}|}, \qquad (10.1)$$

where the centre-of-mass speed v_{CM} is given by:

$$v_{CM} = \frac{m_{Cl}}{m_{C_2H_6} + m_{Cl}} v_{Cl}. \tag{10.2}$$

As explained in detail in Chapter 3 and elsewhere [22], the speed distribution is obtained by integrating the Abel-inverted image over all angles as a function of speed (radius r), and weighing the intensity of each pixel of the image by $r\sin\theta$ (surface element). Once the speed distribution has been determined, with respect to the photolysis centre, the differential cross-section can be obtained by a method similar to that used in photoloc TOF experiments [21]. Instead, we analyse our images as follows: unlike the speed distribution procedure, we integrate the Abel-inverted image as a function of radius *without any surface element weighing*. This integration yields the differential cross-section *directly* (albeit the x-axis is the lab speed $\mathbf{v_{HCl}}$, instead of the scattering angle). Transforming $\mathbf{v_{HCl}}$ to the scattering angle is achieved with (10.1). The resulting HCl($v = 0$, $J = 1$) product-scattering angle distribution [using (10.1)] is shown in Fig. 10.3d. It is interesting that this analysis doesn't require knowledge of the angular distribution (which assumes that there is no internal energy in the ethyl radical) of the hot-atom reactant (Fig. 10.2c). In comparison with the results of Kandel *et al.* that are also shown in Fig. 10.3c, our HCl product angular distribution is more forward scattered. The main difference in the two experiments is the collision energy, which in our case is about 1.5 times greater. This increase in collision energy and the assumption of a stripping-type hydrogen abstraction reaction may explain the discrepancy. A faster approaching Cl atom could strip the H atom with greater ease and at larger impact parameters.

It is important to emphasize that the analysis described above to yield the differential cross-section assumes that no energy is deposited in the internal modes of the unobserved ethyl-radical product. As discussed extensively by Kandel *et al.* [6–8] this assumption can be tested by measuring the product spatial anisotropy parameter β as a function of the lab speed (see Fig. 10.3c). In this case knowledge of the spatial anisotropy of the hot-atom reactant is necessary. Two theoretical curves are shown, demonstrating the two extreme behaviours of β versus lab speed. Curve A describes β when all the available energy to the ethyl radical is kinetic, and curve B describes β when all the available energy is deposited in the internal modes of the ethyl radical. As observed by Kandel *et al.*, the experimental measurements are much closer to curve A than B, indicating that little of the available energy appears in internal modes of the ethyl radical.

10.4 Conclusions

We have demonstrated the potential of applying velocity mapping to measure state-resolved differential cross-sections of bimolecular reactions in a single beam. For the reaction of Cl with C_2H_6 at a collision energy of 0.36 eV our results indicate a forward-scattered $HCl(v = 0, J = 1)$ product, with little internal energy present in the unobserved ethyl radical. These results are similar to previous observations in the same collision energy region.

Acknowledgements

This work is conducted at the Ultraviolet Laser Facility operating at FORTH- (TMR, Access to Large Scale Facilities EU program, Contract No. CHGE-CT92-007) and is also supported by TMR Network *IMAGINE* ERB 4061 PL 97-0264.

References

1. P. Casavecchia, N. Balucani, G. G. Volpi, *Ann. Rev. Phys. Chem.* **50**, 347 (1999).
2. N. E. Shafer, A. J. Orr-Ewing, W. R. Simpson, H. Xu, R. N. Zare, *Chem. Phys. Lett.* **212**, 155 (1993).
3. W. R. Simpson, A. J. Orr-Ewing, T. P. Rakitzis, S. A. Kandel, R. N. Zare, *J. Chem. Phys.* **103**, 7299 (1995).
4. A. J. R. Heck, D. W. Chandler, *Annu. Rev. Phys. Chem.* **46**, 335 (1995).
5. C. R. Gebhardt, T. P. Rakitzis, P. C. Samartzis, V. Ladopoulos, T. N. Kitsopoulos, *Rev. Sci. Instrum.* **72**, 3848 (2001).
6. A. S. Kandel, P. T. Rakitzis, T. Lev-On, R. N. Zare, *J. Chem. Phys.* **105**, 7550 (1996).
7. A. S. Kandel, P. T. Rakitzis, T. Lev-On, R. N. Zare, *Chem. Phys. Lett.* **265**, 121 (1997).
8. A. S. Kandel, P. T. Rakitzis, T. Lev-On, R. N. Zare, *J. Phys. Chem. A* **102**, 2270 (1998).
9. P. C. Samartzis, I. Sakellariou, T. Gougousi, T. N. Kitsopoulos, *J. Chem. Phys.* **107**, 43 (1997).
10. A. T. J. B. Eppink, D. H. Parker, *Rev. Sci. Instrum.* **68**, 3477 (1997).
11. P. C. Samartzis, T. Gougousi, T. N. Kitsopoulos, *Laser Chem.* **17**, 185 (1998).
12. D. S. Green, G. A. Bickel, S. C. Wallace, *J. Mol. Spectrosc.* **150**, 388 (1991).
13. D. S. Green, G. A. Bickel, S. C. Wallace, *J. Mol. Spectrosc.* **150**, 354 (1991).
14. D. S. Green, G. A. Bickel, S. C. Wallace, *J. Mol. Spectrosc.* **150**, 303 (1991).
15. D. W. Chandler, T. N. Kitsopoulos, M. A. Buntine, D. P. Baldwin, R. I. McKay, A. J. R. Heck, R. N. Zare, In: *Gas-Phase Chemical Reaction Systems: Experiments and Models 100 Years after Max Bodenstein*, eds J. Wolfrum, H.-R. Volpp, R. Rannacher, J. Warnatz, Springer Series in Chem. Phys., (Springer, Berlin, Heidelberg, 1996).
16. P. C. Samartzis, I. Sakellariou, T. Gougousi, T. N. Kitsopoulos, *J. Chem. Phys.* **107**, 43 (1997).
17. K. P. Huber, G. Herzberg, *Constants of Diatomic Molecules*, (Van Nostrand Rheinhold Co., New York, 1979), p. 285.
18. P. Marshall, *J. Phys. Chem.*, **103**, 4560–4563 (1999), and references therein.
19. W. J. van der Zande, R. Zhang, R. N. Zare, K. G. McKendrick, J. J. Valentini, *J. Phys. Chem.* **95**, 8205 (1991).
20. R. N. Zare, *Mol. Photochem.* **4**, 1 (1972).
21. N. E. Shafer, A. J. Orr-Ewing, W. R. Simpson, H. Xu, R. N. Zare, *Chem. Phys. Lett.* **212**, 155 (1993).
22. T. N. Kitsopoulos, M. A. Buntine, D. P. Baldwin, R. N. Zare, D. W. Chandler, *Science* **260**, 1605–10 (1993).

11

Slice imaging: a new approach to ion imaging and velocity mapping

T. PETER RAKITZIS AND THEOFANIS N. KITSOPOULOS

11.1 Introduction

Houston and Chandler introduced ion imaging in 1987 [1], demonstrating for the first time the potential of this method to be used in chemical dynamics studies. In most chemical dynamics experiments the desired quantity to measure is the state-resolved differential cross section ($d\sigma/d\Omega$) [2]. This quantity is defined as the amount of product of a chemical reaction, be it a half collision (photodissociation) or a full collision, that is scattered into a unit solid angle per unit of time. Borrowing from the methods of nuclear physics, the pioneers of scattering experiments used the time-of-flight method (TOF) [3] coupled to a rotatable detector to map out $d\sigma/d\Omega$. Initial experiments employed a *universal ionizer* to ionize and subsequently detect the products [3]. The universality of the method made it the most successful method for studying a large number of reactive collisions. However, its limited energy resolution was insufficient for detailed studies of unimolecular processes.

It was quickly realized that using high-resolution laser spectroscopic detection of the products would yield a great deal more information than the universal detection [3]. In order to obtain information concerning $d\sigma/d\Omega$ the most popular methods used were Doppler spectroscopy, TOF, or Doppler coupled with TOF [3]. Ion imaging was introduced as a method that combined Doppler and TOF. Its major drawback was its limiting energy resolution, typically 15–20%, coupled with the 'magical and mysterious' inverse Abel transformation. Eppink and Parker introduced the remedy to the relatively low resolution in 1998 [4]. They used an inhomogeneous extraction field that essentially *focused ion velocities onto the imaging plane,* thus improving the energy resolution by almost an order of magnitude to 2–5%. However, the necessity for the inverse Abel transform remained.

The use of the Abel transform requires an axis of cylindrical symmetry parallel to the imaging plane. For double resonance experiments, for which photofragment polarization is present, both the photolysis and probe laser polarization directions must be parallel if the inverse Abel transform is to be used. While it is possible to measure the photofragment alignment from Abel-invertable images [5], it may be more convenient to do so directly from non-cylindrical symmetric slice images. Most importantly, certain polarization parameters, most notably the $Im[a_1^{(k)}(\parallel, \perp)]$ interference parameter that was introduced in Chapter 4, can *only* be measured from non-Abel-invertable geometries (when the laser polarizations are *not* parallel).

In this chapter we introduce an alternative method to ion imaging that includes features of velocity mapping and allows the *direct* measurement of $d\sigma/d\Omega$, thereby eliminating the need for inversion techniques such as the Abel transform. In addition, the energy resolution remains comparable with that of velocity mapping. By way of an illustrative test case we study the photodissociation of Cl_2 at 355 nm by probing the ground spin-orbit state of the Cl-atom photofragment.

11.2 Basic principles

11.2.1 Kinematics of a single uniform acceleration field

This method has been presented in recent reviews [6,7], however, we would like to go over the kinematics of the ions, primarily along the TOF axis, as this is the essential point in our new approach. A schematic of a traditional ion imaging setup is shown in Fig. 11.1. Ions are created in a homogenous extraction field E that accelerates them towards the position-sensitive imaging detector (PSD). To achieve mass resolution, the extraction field is coupled to a field-free drift region that prolongs the flight time of the ions towards the detector. In addition this drift region allows sufficient time for the ion cloud to expand in diameter thus achieving better spatial resolution (i.e. better energy resolution).

For the setup of Fig. 11.1, the general TOF T for an ion of mass m and charge state q with initial velocity component v_0 along the TOF axis, is given by

$$T(d, v_0) = \frac{1}{a} \left[\sqrt{v_0^2 + 2ad} - v_0 \right] + \frac{L}{\sqrt{v_0^2 + 2ad}}, \qquad (11.1)$$

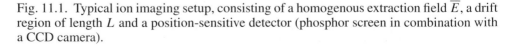

Fig. 11.1. Typical ion imaging setup, consisting of a homogenous extraction field $\overline{E}$, a drift region of length L and a position-sensitive detector (phosphor screen in combination with a CCD camera).

where $a = (qE)/m$ is the acceleration of the extraction field, and d is the position with respect to the extractor grid.

The initial particle velocity v_0 is a signed value. It is positive for 'forward scattered particles', those *initially* moving towards the detector, and negative for 'backward scattered particles', those *initially* moving away from the detector. In the case of photodissociation, v_0 becomes

$$v_0 = c + u_0 \cos(\alpha) \tag{11.2}$$

where c is the molecular beam velocity, assumed to be parallel to the TOF axis, u_0 is the velocity of the photofragments and α is the angle between the direction of the photofragments and the TOF axis. The component $u_0 \sin(\alpha)$ of the photofragment velocity is what controls the dimension of the ion cloud perpendicular to the TOF axis. Neglecting the relatively small time that ions spend in the extraction region (acceleration time)[1], the diameter of the ion image on the detector is given by

$$\Delta y = \frac{2u_0 L}{\sqrt{c^2 + 2ad}}. \tag{11.3}$$

Thus, the minimum acceleration needed to image the ion cloud onto a detector of diameter D is given by the relationship

$$a = \frac{1}{2d} \left[\left(\frac{2u_0 L}{D} \right)^2 - c^2 \right]. \tag{11.4}$$

Figure 11.2a shows a simulation of a photofragment imaging experiment following the photodissociation of a diatomic molecule ($A_2 \rightarrow 2A$). The parameters have been chosen to match our experimental conditions: $L = 45$ cm, $d = 1$ cm, $E = 333$ V/cm, $c = 1240$ m/s, $u_0 = 1668$ m/s. The spatial extent of the ion cloud is shown at several time intervals. It is immediately obvious that although the vertical dimension of the ion cloud (Δy) increases considerably during its drift time towards the detector, the horizontal spread (Δx) is nearly invariant. In fact, in the limit of zero beam velocity Δx in the drift region is a *constant* given by $u_0 \Delta T_r$, where $\Delta T_r = 2v_0/a$ is the 'turn-around-time' of the backward scattered ions in the extraction region[2].

In Fig. 11.2b we plot the temporal spread of the ion packet along the TOF axis as a function of u_0, for the conditions used in the simulation. We observe that ΔT is *extremely* small and varies almost linearly with u_0.

---

[1] For typical velocities (10^5 m/s) and moderate extraction field strengths (~ 100 V/cm), $T(d, v)$ becomes approximately velocity independent, i.e. $T(d, v) \approx T(d)$.

[2] For the backscattered particle, when $u_0 > c$, the particle begins to move towards the repeller, it is decelerated until its velocity becomes zero, and then starts to accelerate in the forward direction towards the detector.

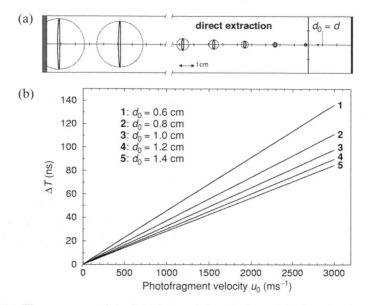

Fig. 11.2. (a) The upper panel depicts the spatial evolution of the ion cloud on its way to the detector for the case of a *DC extraction field*. The simulation has been performed for ground-state Cl from the photodissociation of Cl_2 at 355 nm ($c = 1240$ m/s, $u_0 = 1700$ m/s, $L = 45$ cm, $d_0 = d_i = 1$ cm, $E = 333$ V/cm). The thick line represents the position of the ions at 0, 0.25, 0.50, 1, 1.5, 2.5, 9.75 and 10.75 μs, assuming for simplicity an isotropic photofragment distribution. The dotted circle is given to emphasize the distortion of the ion cloud by the acceleration. (b) The lower panel shows the time width ΔT of the ion cloud at the detector as a function of the photofragment velocity. Here, a molecular beam velocity $c = 1000$ m/s has been assumed. The different curves correspond to different photodissociation and post-ionization positions d_0 in the extraction region. Similar curves can be obtained by varying the extraction field E, which is equivalent to varying d_0.

11.2.2 Conventional versus pulsed extraction (slice imaging)

As we saw in the previous paragraph, when using constant (DC) extraction fields, there is always a *crushing* of the ion Newton sphere along the TOF axis. By introducing a time delay τ following the photodissociation and post-ionization processes, the position of the ions with respect to the extractor grid are

$$d = d_0 - c\tau \tag{11.5a}$$

$$d_f = d_0 - (c + u_0)\tau = d - u_0\tau \tag{11.5b}$$

$$d_b = d_0 - (c - u_0)\tau = d + u_0\tau, \tag{11.5c}$$

where d_0 is the initial creation position of the ions ($\tau = 0$); d_f is the position of the forward-scattered ions ($\alpha = 0$) and d_b is the position of the backward-scattered ions ($\alpha = \pi$) after time τ, when the extraction field is pulsed on. In this case, the

(a)

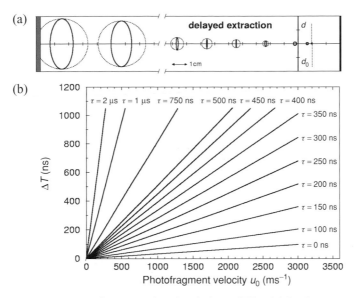

Fig. 11.3. (a) The upper panel repeats the simulation of Fig. 11.2a; however, in this case for a *delayed extraction* of the ion cloud ($\tau = 250$ ns). A space focus appears in the drift region after which the axial width of the ion cloud increases rapidly. The lower panel shows the time width of the ion cloud at the detector as a function of the photofragment velocity u_0 ($c = 1000$ m/s). The different curves correspond to different delay times τ. ΔT increases almost linearly over a wide range of photofragment velocities as well as extraction delays. The dependency of ΔT on τ shows a minimum at zero delay times (DC extraction).

temporal spread of the ion cloud along the TOF axis becomes

$$\Delta T = |T_f(d_f, v_f) - T_b(d_b, v_b)| \tag{11.6}$$

The influence of the extraction delay τ on the spatial evolution of the ion cloud is shown in Fig.11.3a. The parameters in this simulation are identical to those used in Fig. 11.2a except that the photodissociation and post-ionization take place in field-free conditions, i.e., both the repeller and extractor grid have equal potentials (typically grounded). After a delay of 250 ns the repeller is switched rapidly to 1 kV, so that the ions experience an extraction field of 333 V/cm. From the evolution of the width Δx of the ion cloud shown in Fig. 11.3a we observe the following. During the time delay τ, the ion cloud expands to some width (~ 0.8 mm), then, following the application of the extraction field, Δx decreases, then vanishes at the '*geometric* focus' [8] (located a few centimetres after the extractor grid) and subsequently increases rapidly. This increased spatial width naturally corresponds to some larger spread in the arrival times of the ions. The temporal width ΔT is plotted in Fig 11.3b as a function of velocity u_0 for various extraction delays. This delay extraction time thus provides a handle on controlling the size of the ion packet along the TOF axis, *while in the meantime ensuring that the perpendicular size of the packet fits on the imaging detector.*

This technique is analogous to the *time lag* focusing introduced by Wiley and McLaren in the seminal paper on TOF mass spectrometry [8]. In their case however, the time lag was introduced in order to improve mass resolution and thus intended to reduce the temporal spread of the ion cloud caused by the *initial* velocity spread. This was achieved by choosing the operating conditions such that the spatial focus of the ion cloud occurred at the detector. In our case, it is desired to operate far from the spatial focus such that the maximal spread in the ion cloud is achieved.

Figure 11.3b indicates that temporal spread of the order of 400 ns can be achieved readily even for extremely low speeds ($u_0 \approx 100$ m/s). Using a moderately fast time gate (< 40 ns) on our imaging detector, we are able to image only the *central slice* of the ion cloud that is equivalent to an intensity slice through the three-dimensional photofragment distribution. This image (in the limit of infinitesimal slicing) is equivalent to the result obtained by using the inverse Abel transform in conventional imaging.

11.3 Experimental

The apparatus has been described in detail elsewhere [9]. Briefly, a gas sample containing 5% Cl_2 in He is expanded into the source vacuum chamber via a home-built piezoelectrically actuated pulsed molecular beam operating at 10 Hz. After passing through a skimmer and collimator, the molecular beam is intersected at right angles by two counter propagating laser beams. The photolysis laser beam is generated by the third harmonic of a Nd:YAG laser (BMI series 5000). The probe laser beam is generated by a MOPO-SL laser (Spectra Physics 730DT10). The Cl-atom photofragments are ionized using two-photon resonant transitions $4p(^4P_{3/2}) \leftarrow 3p^5(^2P_{3/2})$ 240.53 nm [10–12] and the ions produced are accelerated towards a home-built ion-imaging detector. Ions of different mass separate in their time-of-flight during their field-free trajectory on route to the detector. The length of this field-free drift region is ~ 45 cm. The detector gain is pulsed ON at the proper arrival time for mass-selection. Images appearing on the detector anode are recorded using a CCD camera (Spectra Source).

11.3.1 Ion optics

A schematic of the ion optics used in this experiment is shown in Fig. 11.4. The extraction field consists of the repeller electrode that also serves as the molecular beam collimator and the extractor, a grounded plate consisting of a flat grid. The quality of the grid is critical as the aberrations introduced by the grid are on the order of the grid hole dimension. Hence it is desirable for the grid hole size to be

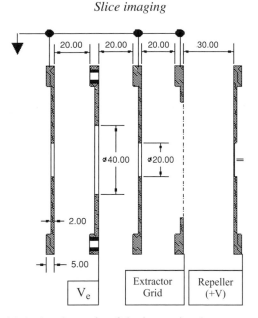

Fig. 11.4. A schematic of the ion optics drawn to scale.

on the order of 50 to 100 μm (spatial resolution of imaging detector). For the results presented here the grid used is an electroformed mesh with 333 lines per inch (Buckbey Mears MN-38 BM-0333-01), hole size 58 μm, wire thickness 12 μm. The separation between repeller and extractor is 3 cm while the lasers cross the molecular beam approximately 1 cm from the extractor grid. An Einzel lens assembly is mounted after the extractor that consists of three slit electrodes separated by 2 cm. The two outer-grounded electrodes of the Einzel have hole diameters of 2 cm while the middle electrode has a hole diameter of 4 cm. The Einzel lens is used to *velocity map* the photofragments thus correcting for the finite extent of the interaction volume. We find that the lens operating voltage is not as critical as in the case of conventional velocity mapping geometry described in the pioneering paper of Eppink and Parker [4]. We observe that beyond a *minimum* velocity mapping voltage, the lens operates as a geometric focusing lens thus allowing one to zoom in or out of desired parts of the image without affecting the spatial resolution. For the present geometry we find that for a repeller voltage of 1000 V an Einzel lens voltage of ∼ 150 V is sufficient for velocity mapping to be achieved, and these voltages scale approximately linearly with the repeller voltage.

11.3.2 Repeller high voltage switch

The circuit of the high voltage switch used for pulsing the repeller electrode is shown in Fig. 11.5. The repeller voltage generator is a high voltage pulser that is a

High voltage repeller pulser

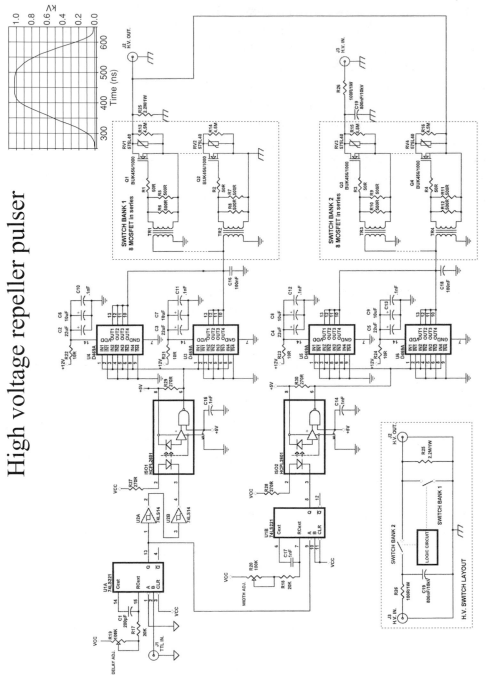

Fig. 11.5. The circuit of the high voltage pulser used for pulsing the repeller electrode. The insert demonstrates the performance of the pulser when loaded with the ion optics assembly.

modification of a previously published circuit [13], with the addition of a grounding switch bank 1 at the output, and the appropriate driving circuitry. Each bank consists of eight high voltage MOSFET switches, thus allowing the pulser to deliver pulses of more than 5 kV in amplitude. The addition of switch bank 1 is used to make the switch OFF times faster. A trigger pulse with the appropriate timing arrives at input J1 and in turn causes the one shot U1A to fire. The ON time duration is controlled by R19. This pulse controls the ON time of switch bank 2, which is the pass switch. At the falling edge of this pulse U1B fires, which in turn drives switch bank 1 ON and the output returns to 0 voltage. This is done so that the turn OFF time of the pulser is as fast as the turn ON time, giving it equal rise and fall times. The performance of the pulser in switching the repeller voltage is shown in the insert of Fig. 11.5. The rise/fall time of $\sim$80 ns (10–90% max) can be achieved.

11.3.3 Detector gating

The imaging detector consists of a pair of matched 4 cm diameter microchannel plates (HAMAMATSU) coupled to a phosphor screen (P46 Proxitronic). To gate the gain of the detector the voltage on the back MCP is set to $\sim$ 1100 V and the front MCP is pulsed to -500 V at the proper timing. The circuit of the fast high voltage switch used for the pulsing is shown in Fig. 11.6. Input J3 receives the gating trigger which fires the one shot U1A at the rising edge. This pulse drives the gate of switch Q1 which turns ON very fast, causing the voltage at its drain to go from 350 V to zero. When the driving pulse returns to zero, the switching transistor turns OFF quickly due to reverse voltage characteristics of the isolating transformer TR1. Capacitor C10 and diode D1 shift the output voltage so that it swings between ground and $-V$. The pulse width is determined by R3, and the minimum pulse duration which enables sufficient gain for single particle detection is $\sim$ 40 ns and the performance of the switch is shown in the insert of Fig. 11.6. As will be shown later, the actual time window achieved with this switch is $\sim$ 25 ns as only the central-most part of the pulse contributes sufficient gain to the MCP stack.

11.4 Results

The photodissociation of Cl_2 at 355 m has been extensively studied. The primary reaction channel is

$$Cl_2 \rightarrow 2Cl(^2P_{3/2}) \tag{11.7}$$

with a branching ratio of greater that 99% [14,15]. The transition is a perpendicular one as the molecule is excited from a $X(0_u^+, \Omega = 0)$ ground state to the $C(1_u, \Omega = 1)$ state. This means that the transition dipole for this excitation lies perpendicular to

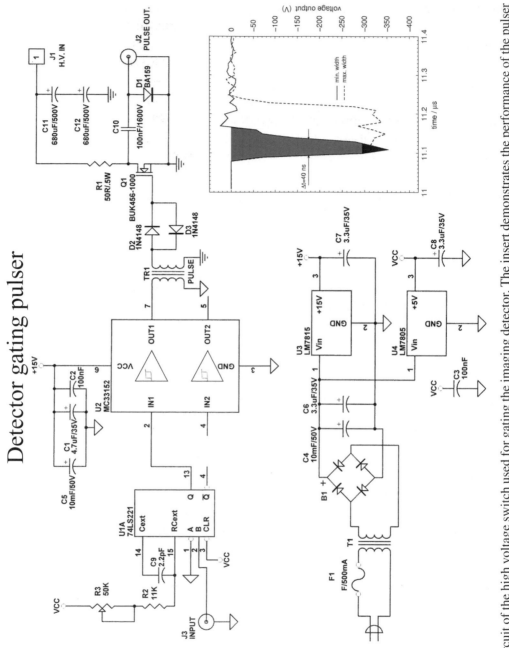

Fig. 11.6. The circuit of the high voltage switch used for gating the imaging detector. The insert demonstrates the performance of the pulser when connected to the detector. While the minimum pulse width is 40 ns nominally, the effective pulse width is on the order of 25 ns (black tip of the pulse) due to the nonlinear amplification characteristic of the MCP stack.

the molecular bond axis such that a molecule will have the maximum absorption probability when it lies perpendicular to the laser polarization (electric field direction) [16].

11.4.1 Slice imaging

The characteristics of slice imaging are summarized in Fig. 11.7. The region of interest of the mass TOF spectrum is shown in Fig. 11.7a. For $\tau = 0$ the mass spectrum consists of two narrow peaks which correspond to ^{35}Cl and ^{37}Cl.

As predicted by the simulation in Fig. 11.3a, increasing the time delay to 250 ns causes the two peaks to broaden and start to be partially resolved. Further increase in the time delay yields a broad unresolved mass peak. Hence although increasing the time delay would allow more facile *slicing*, when multiple isotopes of the species being probed are present, the upper limit in the time delay is set by the amount of overlap between the isotope peaks. We wish to point out here that even in the absence of any isotopes, if the delay times become too large, then the dimensions of the ion cloud become comparable to the ion optic dimensions. When this happens, distortions due to edge effects become apparent.

A series of images are presented in Fig. 11.7 that are acquired at 25 ns time intervals indicated by the comb structure on Fig. 11.7a, thus slicing through the ^{35}Cl ion cloud for an extraction delay $\tau = 250$ ns. The polarization of the photolysis laser for all the images presented here is linear and vertical with respect to the figure frame. Image A is the fastest part of the ion cloud and corresponds to fragments that are initially backscattered (as defined above). The appearance of a bright central core and a relatively small radius is what one expects for a sphere with an anisotropy parameter of $\beta = -1$ [15,16], i.e., maximum intensity along the equator. The faint signal at larger radii is some leakage of the entire Cl ion cloud because the residual 1100 DC voltage on the MCP stack is sufficient to just detect some ions. At longer arrival times, the radius of the image increases and becomes brighter on the edge and hollow in the centre. *The image radius becomes maximum for image H. This image corresponds to the centre slice.* The reverse trend follows as we continue to probe at even longer arrival times, and eventually we start *slicing* mass ^{37}Cl. Thus we conclude that a time delay of $\tau = 250$ ns is sufficient for slice imaging the ^{35}Cl photofragments.

11.4.2 Conventional velocity mapping versus slice imaging

Figure 11.8a shows a typical ^{35}Cl photofragment image obtained using conventional velocity mapping, and the appearance of this image is consistent with the predictions above ($\beta = -1$) and its detailed analysis has been presented elsewhere [15]. After

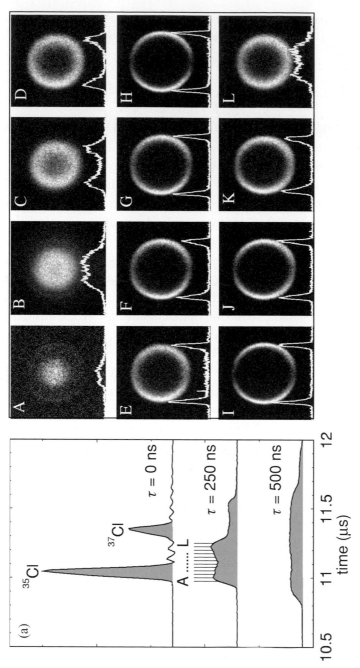

Fig. 11.7. (a) The left panel depicts time-of-flight mass spectra for chlorine ions for various extraction delays τ. (b) With increasing τ the peaks broaden considerably in time. For a time delay of $\tau = 250$ ns images have been taken for individual slices of the ion cloud at the positions indicated by A to L. These images are given in the right panel. Each of them has been normalized to increase contrast for visualization. The relative signal intensities are contained in the overlaid curves, which represent horizontal cuts through the centre of the image (summed over five adjacent rows).

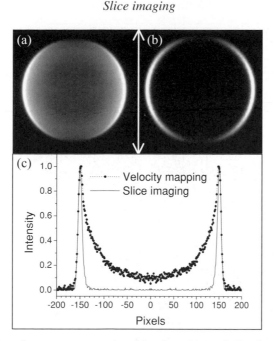

Fig. 11.8. Comparison between conventional ion imaging and slice imaging. (a) Reprodc-tion of a typical ^{35}Cl photofragment image (see Ref. [15]) obtained under velocity mapping conditions. (b) Image obtained using the slice imaging technique. It corresponds to image H in Fig. 11.7b. The arrow indicates the polarization axis of the photodissociation and ion-ization lasers. (c) The lower panel compares intensity profiles for cuts along the equator of the two images above.

the appropriate image processing the speed and angular distributions are determined and presented in Fig. 11.9 and Fig. 11.10 [6,17]. Figure 11.8b is the *centre slice image* H [Fig. 11.7b].

The intensity cuts along the equator of the two images are presented in Fig. 11.8c. For conventional velocity mapping the profile is characteristic of $\beta = -1$ [16], and one major advantage of slice imaging becomes immediately apparent. The long tail inherent in the projection of the entire photofragment Newton sphere has been eliminated when performing slice imaging. The sharpness of the outer edges are identical, thus indicating that the grid used in the extractor is of sufficient quality such that it does not degrade the resolution of the image obtained, at least in com-parison to conventional velocity mapping. Considering a situation where multiple channels are involved, a very weak interior peak can be hidden under the tail of the outer peak and will thus be missed when performing the inverse Abel transform. In the case of slice imaging, however, such a situation is circumvented.

Recently, a further improvement in velocity mapping has been the use of an ion event counting acquisition method [18,19], whereby during the acquisition one centroids each ion event (blob/spot), sets the intensity of the *centroid pixel* to

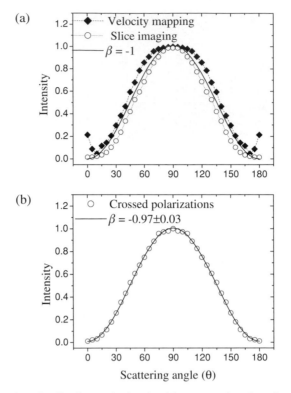

Fig. 11.9. (a) Angular distributions obtained with conventional and with slice imaging when the polarizations of the photolysis and probe laser are parallel. Also shown is a fit to both data for $\beta = -1$. (b) Angular distributions obtained with slice imaging when the polarizations of the photolysis and probe lasers are perpendicular.

one (1) and all the other pixels to zero. Although this is an excellent way to correct for detector inhomogeneities it takes a very long time to acquire enough *centroid statistics* to fill the entire image and thus enable one to perform the inverse Abel transform. The usual practice is to *blur* the pixilated image with some Gaussian function (typically 3×3 pixels) and subsequently perform the inverse Abel transform. This procedure however can be dangerous as the inverse Abel transform requires that *the entire phase space of the photofragment Newton sphere be sampled, and this is NOT the same as smoothing sharp features using a Gaussian function*. If however one were to apply the event counting procedure to slice imaging there are at least two advantages. As no mathematical transformation will be applied, there is no need for smoothing of the data using a Gaussian function, so that the resolution is not artificially degraded. Since only the centre slice of the entire Newton sphere is detected, there will be a much smaller number of *overlapping multiple ion events that lead to false centres*. Hence, event counting might

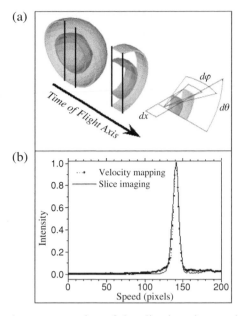

Fig. 11.10. (a) Schematic representation of the slice imaging method. The central part of two Newton spheres is cut out by means of a gated detector. The images obtained in this way are a projection of a very finite strip of the Newton sphere. The lower panel compares the speed distributions obtained with velocity mapping and with slice imaging. The curves correspond to the images given in Fig. 11.8.

possibly improve the resolution of slice imaging with respect to the work presented here.

11.4.3 Extracting angular distributions from slice images

The angular distribution $I(\theta)$ for *each speed* appearing in *slice image* is obtained by integrating the image in the same manner as in the case of the inverse Abel transformed image [17]. That is

$$I(\theta) = \frac{N(\theta)}{\Delta\theta} \tag{11.8}$$

where $N(\theta)$ is the total pixel intensity in a radial sector centred at the centre of mass of the image, with an angle $\Delta\theta$.

Three distinct advantages of the slice imaging technique are:

(a) The noise introduced by the inverse Abel transform, particularly along the symmetry axis, is eliminated. This feature is demonstrated in Fig. 11.9a where the angular distribution obtained using the inverse Abel transform shows artificial noise near the symmetry axis, while this noise is *absent* in the angular distribution determined from slice imaging.

(b) Abel invertible geometries are no longer a prerequisite. This is a tremendous asset when photofragment alignment and orientation are present. For the case in hand, it has been shown that the Cl photofragments are aligned [20–22]. This leads to deviations from the limiting β value of -1 when the photolysis and probe polarizations are parallel, as is evident in both angular distributions in Fig. 11.9a. However, if one rotates the probe polarization $90°$ with respect to the imaging plane (such that the angle between the probe polarization and all the photofragment velocities detected in the slice image is approximately $90°$), then there is no variation in the angular distribution due to alignment effects, and the true β value should be measured. This is confirmed in Fig. 11.9b where the fit to the data yields $\beta = -0.97 \pm 0.03$.

(c) It is no longer necessary to have all velocities of the photofragmentation process on the *same image* as the inverse Abel transform is no longer needed. Hence one can *expand* the image on the detector as much as desired (thereby losing the fast components) without any loss of information for the slow components. In addition, the use of uniform electric field makes it possible to operate even at very low repeller voltages (< 100 V), which is not possible with conventional velocity mapping because of the lens aberrations.

11.4.4 Extracting speed distributions from slice images

Unlike the inverse Abel transform which yields an image representing the intensity slice along a plane containing the symmetry axis, slice images are actually a projection of a very thin finite stripe of the Newton sphere. In order to obtain the speed distribution with the usual $\rho\sin\theta$ weighing of each pixel [17], we must first divide out by the *azimuthal contribution to the solid angle subtended at each radius*. Figure 11.10a shows the slicing of two concentric Newton spheres. Assuming that the lab. velocity for the near perpendicular direction of each concentric ion cloud is about the same, then the *spatial extent (dx)* of the slice for each Newton sphere along the TOF axis will be approximately constant for all radii. If the photolysis laser polarization is parallel to the *column direction* of the data image, then for *a row of data*, the *slice* intensity $I_R(\theta)$ can be related to corresponding intensity obtained from the conventional inverse Abel transform $I_R^A(\rho)$ as follows:

$$I_R(\rho) \approx I_R^A(\rho)dx \Rightarrow I_R^A(\rho) \propto I_R(\rho). \tag{11.9}$$

Hence we observe that the procedure for obtaining the speed distribution for slice images is identical to the procedure used for conventional ion imaging and/or velocity mapping, i.e., we weight each pixel by $\rho\sin\theta$.

The speed distributions for the Cl atoms determined using velocity mapping and slice imaging are presented in Fig. 11.10b. The widths of the two distributions are comparable. However, as we explained earlier, we believe that the *effective* resolution of slice imaging is in principle better than that of conventional imaging

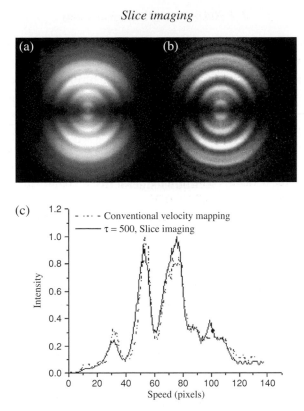

Fig. 11.11. (a) Conventional velocity map image for the (2+1) REMPI of Cl_2 via the $^1\Pi_{1g}(4s)(v = 1)$ state (Ref. [23]). (b) Slice image for the (2+1) REMPI of Cl_2 via the $^1\Pi_{1g}(4s)(v = 1)$ state. (c) Corresponding speed distributions for images (a) and (b).

as the tail in the intensity profile of the image projection is not present in the slice image. One can of course perform peak shaping to the intensity profile of the slice image to enhance even more the resolution, but the beauty of this new method lies in its direct determination of both speed and angular distribution without any mathematical transformation.

In Fig. 11.11 we present the images and corresponding speed distributions for Cl^+ ions produced from the (2+1) REMPI of Cl_2 via the $^1\Pi_{1g}(4s)(v = 1)$ state, measured both by the velocity mapping and slice imaging methods [23]. The agreement is very good, confirming that (11.9) above is valid and that slice imaging can be used for quantitative determination of branching ratios.

11.5 Conclusions

We have demonstrated a new approach to measuring state-resolved velocity distributions of photofragments. The method allows the direct determinations of the scattering information for the slice image without the need for any image processing such as the inverse Abel transform implemented in conventional imaging approaches. This method is completely general and can be applied to photoelectron imaging and reaction product imaging. In the case of photoelectron imaging *extra caution* must be taken to eliminate stray fields (both magnetic and electric) from the interaction region, as this will perturb the electron trajectories during 'field free' extraction delay.

Acknowledgements

We thank Mr. Giorgos Gousis for outstanding technical support in building the high voltage pulsers. This work was conducted at the Ultraviolet Laser Facility operating at FORTH-IESL (Improving Human Potential-Transnational Access to Major Research Infrastructures, HPRI-CT-1999-00074) and is also supported by TMR Network *IMAGINE* ERB 4061 PL 97–02.

References

1. D. W. Chandler, P. L. Houston, *J. Chem. Phys.* **87**, 1445 (1987).
2. R. D. Levine, R. B. Bernstein, *Molecular Reaction Dynamics and Chemical Reactivity*, (Oxford, Oxford University Press, 1987).
3. *Atomic and Molecular Beam Methods Vols I and II*, ed. G. Scoles, (New York, Oxford, Oxford University Press, 1992).
4. A. T. J. B. Eppink, D. H. Parker, *Rev. Sci. Instrum.* **68**, 3477 (1997).
5. T. P. Rakitzis, P. C. Samartzis, T. N. Kitsopoulos, *Phys. Rev. Lett.* **87**, 3001 (2001).
6. A. J. R. Heck, D. W. Chandler, *Annu. Rev. Phys. Chem.* **46**, 335 (1995).
7. A. G. Suits, R. E. Continetti, *Imaging in Chemical Dynamics* (Washington, ACS Symposium Series 770, 2001).
8. W. C. Wiley, I. H. McLaren, *Rev. Sci. Instrum.* **26**, 1150 (1955).
9. P. C. Samartzis, I. Sakellariou, T. Gougousi, T. N. Kitsopoulos, *J. Chem. Phys.* **107**, 43 (1997).
10. S. Arepalli, N. Presser, D. Robie, R. J. Gordon, *Chem. Phys. Lett.* **118**, 88 (1985).
11. C. E. Moore, *Atomic Energy Levels as Derived from the Analyses of Optical Spectra*, (Washington, US Dept. of Commerce, National Bureau of 1949).
12. W. R. Simpson, T. P. Rakitzis, A. A. Kandel, J. Orr-Ewing, R. N. Zare, *J. Chem. Phys.* **103**, 7313 (1995).
13. R. E. Continetti, D. R. Cyr, D. M. Neumark, *Rev. Sci. Instrum.* **63**, 1840 (1992).
14. Y. Matsumi, K. Tonokura, M. Kawasaki, *J. Chem. Phys.* **97**, 1065 (1992).
15. P. C. Samartzis, B. L. G. Bakker, T. P. Rakitzis, D. H. Parker, T. N. Kitsopoulos, *J. Chem. Phys.* **110**, 5201 (1999).
16. R. N. Zare, *Mol. Photochem.* **4**, 1 (1972).
17. D. W. Chandler, T. N. Kitsopoulos, M. A. Buntine, D. P. Baldwin, R. I. McKay, A. J. R. Heck, R. N. Zare. In *Gas-Phase Chemical Reaction Systems: Experiments and Models 100 Years after Max Bodenstein*, eds J. Wolfrum, H.-R. Volpp, R. Rannacher, J. Warnatz, (Springer Series in Chem. Phys.; Springer, Berlin, Heidelberg, 1996).
18. L. J. Rogers, M. N. R. Ashfold, Y. Matsumi, M. Kawasaki, B. J. Whitaker, *Chem. Phys. Lett.* **258**, 159 (1996).

19. B. Y. Chang, R. C. Hoetzlein, J. A. Mueller, J. D. Geiser, P. L. Houston, *Rev. Sci. Instrum.* **69**, 1665 (1998).
20. Y. Wang, H. P. Loock, J. Cao, C. X. W. Qian, *J. Chem. Phys.* **102**, 808 (1995).
21. A. S. Bracker, E. R. Wouters, A. G. Suits, Y. T. Lee, O. S. Vasyutinskii, *Phys. Rev. Lett.* **80**, 1626 (1998).
22. T. P. Rakitzis, A. A. Kandel, A. J. Alexander, Z. H. Kim, R. N. Zare, *J. Chem. Phys.* **110**, 3151 (1999).
23. D. H. Parker, B. L. G. Bakker, P. C. Samartzis, T. N. Kitsopoulos, *J. Chem. Phys.* **115**, 1205 (2001).

Index

Bold indicates the first page of a range.

DATE DUE
